# International Conference on Aerospace and Aviation (ICASA 2018)



## Aerospace and Aviation for Regional Sustainable Development

IOP Conference Series: Materials Science and Engineering
Volume 645

Bandung City, Indonesia
23 - 25 October 2018

**Editors:**

**Taufiq Mulyanto**
**Irza Sukmana**
**Khairul Ummah**

ISBN: 978-1-5108-9844-8
ISSN: 1757-8981

**Printed from e-media with permission by:**

Curran Associates, Inc.
57 Morehouse Lane
Red Hook, NY 12571

**Some format issues inherent in the e-media version may also appear in this print version.**



Institute of Physics
Dirac House, Temple Back
Bristol BS1 6BE UK

Phone: 44 1 17 929 7481
Fax: 44 1 17 920 0979

techtracking@iop.org

**Additional copies of this publication are available from:**

Curran Associates, Inc.
57 Morehouse Lane
Red Hook, NY 12571 USA
Phone: 845-758-0400
Fax: 845-758-2633
Email: curran@proceedings.com
Web: www.proceedings.com

# Table of contents

## Volume 645

**The International Conference on Aerospace and Aviation**

**23–25 October 2018, Bandung City, Indonesia**

**Accepted papers received: 23 September 2019**
**Published online: 28 October 2019**

### Preface



### Papers

#### Mechanical and Aerospace Engineering

IOP Conf. Series: Materials Science and Engineering **645** (2019) 011001 doi:10.1088/1757-899X/645/1/011001

# PREFACE

The International Conference on Aerospace and Aviation (ICASA 2018) was organized by and held at Institut Teknologi Bandung, Indonesia on October 23rd – 25th 2018. Highlighting on the current growth in aerospace field and aviation in South East Asia, ICASA 2018 conference themes was taken to be "Aerospace and Aviation for Regional Sustainable Development".

ICASA 2018 accomplished as a forum to exchange and to discuss information in the most advanced scientific fields related to aerospace and aviation and in recent cutting-edge technologies among participants from the world's academia, industries, and governments. The Conference also provides a platform for collaborations among professional societies and to enhance technical exchanges among participants.

We are committed on increasing the quality of the papers through a fruitful peer-review based on the originality, impact, novelty, and the relevance to the area of the conference. The published papers were selected from ICASA 2018 presented paper.

The committee would like to thank you all distinguished keynote speakers, authors, and reviewers for their contribution that have made this conference a forum of exchange, enlightening and network. Also, for all steering committee and organizing committee member and staff that plan and make the conference possible. In addition, ICASA 2018 is launched in a good collaboration with Faculty of Engineering, Universitas Lampung and Java Scientific Academy, Indonesia. We do hope that you would find this publication as an enjoyable and useful guide and reference.

Editorial board:

Taufiq Mulyanto, Faculty of Mechanical and Aerospace Engineering, Institut Teknologi Bandung Indonesia

Irza Sukmana, Faculty of Engineering, Universitas Lampung, Indonesia

Khairul Ummah, Faculty of Mechanical and Aerospace Engineering, Institut Teknologi Bandung Indonesia

IOP Conf. Series: Materials Science and Engineering **645** (2019) 011002 doi:10.1088/1757-899X/645/1/011002

# Peer review statement

All papers published in this volume of *IOP Conference Series: Materials Science and Engineering* have been peer reviewed through processes administered by the proceedings Editors. Reviews were conducted by expert referees to the professional and scientific standards expected of a proceedings journal published by IOP Publishing.

# Development of Star Pattern Recognition Algorithm for Star Sensor Using $C_3^n$ Combination of Triangles Pattern

**N S Ardi[1], R E Poetro[1], M A Saifudin[2], L Fathurrohim[1]**

[1] Faculty of Mechanical and Aerospace Engineering, Bandung Institute of Technology, Indonesia.
[2] Indonesian National Institute of Aeronautics And Space (LAPAN)

E-mail: nugrahaardi53@gmail.com

**Abstract.** The use of star sensor, which is the most accurate attitude sensor on satellites, began to penetrate into micro satellites and nano satellites. Thus the need arises to research and develop star sensor independently to meet the specific design of micro satellites and nano satellites. There are mainly two aspects to be considered in star sensor: hardware and software (algorithm). Star sensor's algorithm is also known as star pattern recognition algorithm. This paper deals with development of star pattern recognition algorithm using $C_3^n$ combination of triangles pattern, that is when there are $n$ stars to be used to do pattern recognition, then $C_3^n$ combination of triangles pattern would be generated. Success rate test are conducted in digital simulation approach using various number of stars that be used to do pattern recognition and various of star image resolution to test its robustness against star spot centroiding error. Results show that this algorithm has high level of success rate even on low resolution of star image, but has increased processing time when more stars are used to do pattern recognition.

**Keyword:** attitude sensor, star sensor, star pattern recognition.

## 1. Introduction

Star sensor is the most accurate attitude sensor in satellite [1], hence it is the most researched attitude sensor among others such as sun sensor, magnetometer, horizon sensor, etc [2]. It has two aspects, namely hardware and software (algorithm) [3]. Hardware aspect consists of buffle to minimize disturbance light, optical lens to focus the image, and imager or image sensor to capture the star image [2]. Algorithm aspect consists of star spot centroiding algorithm to detect position of star's center [4], star pattern recognition algorithm to match pattern from star image with GSC (Guide Star Catalogue) [5], and attitude estimation algorithm to estimate the attitude of star sensor [6]. In algorithm aspect, star pattern recognition algorithm is widely researched because it plays important role: Firstly, a good star sensor should be able to accurately match star image's pattern with star pattern from GSC [7]. If matching process failed, then star sensor failed. Secondly, a good star sensor should has fast processing time in matching star image's pattern with star pattern from GSC to provide first attitude [8]. The simplest method for star pattern recognition is geometric approach using three stars. In this paper, the development of star pattern recognition using $C_3^n$ combination of triangle patterns is presented. The triangle combination is subjected to perform the accuracy. The scope of this work is focused on star pattern recognition and attitude calculation is not presented. The algorithm is performed by digital simulation. The performance of the algorithm is measured by success rate of pattern matching and its processing time for various star image resolutions.

IOP Conf. Series: Materials Science and Engineering **645** (2019) 012001 doi:10.1088/1757-899X/645/1/012001

## 2. Methodology

The methodology used in this paper is digital simulation. Digital simulation is an approach in star sensor research which all of its processes are done in a software [9]. The flowchart of digital simulation is depicted by figure 1

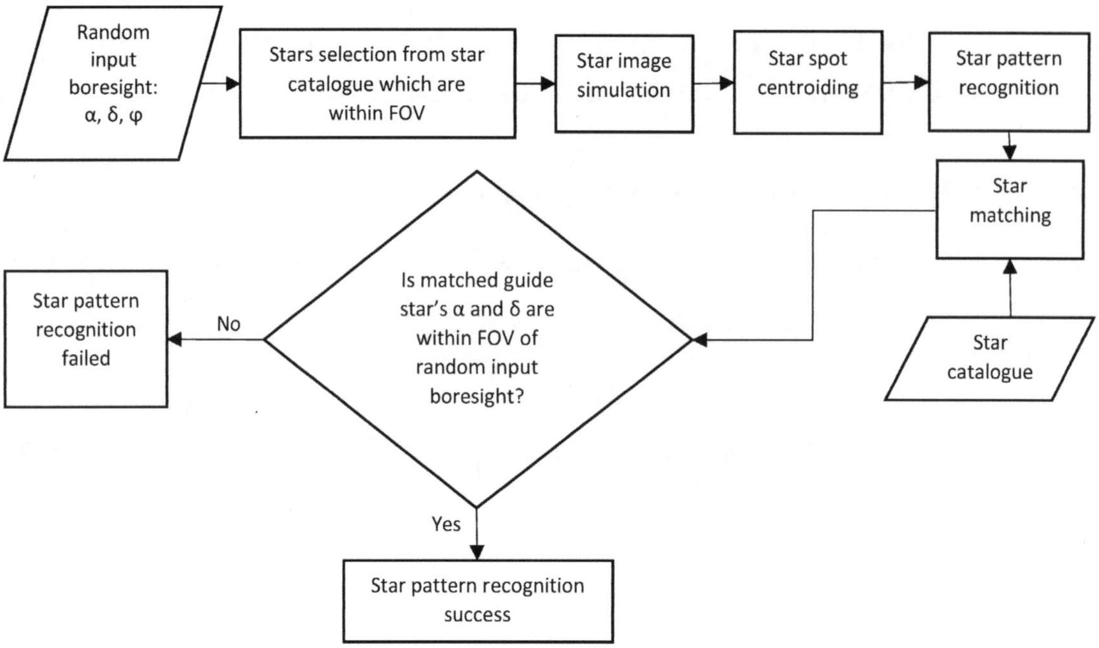


**Figure 1.** Flowchart of digital simulation.

Firstly, random input boresight consists of right ascension ($\alpha$), declination ($\delta$), and roll ($\varphi$) is used to create star image from star catalogue. We use SAO J2000 Star Catalogue with star whose magnitude is less than 6 to simulate the star image. Then the star image is processed to get star's center position using a star spot centroiding algorithm. In this paper we use centroid method [9]. Then star pattern are generated from star image and matcing process is done. If the matched pattern is within FOV, then pattern recognition process is success, and if not, then it is failed. Those processes are done 1000 times for every different boresight to represent all sky direction. Then success rate is calculated by dividing the number of success by 1000. Digital simulation is done for various star image resolutions.

## 3. Star Pattern Recognition Algorithm

The main idea of this algorithm is using $C_3^n$ combination of triangles pattern, that is when there are n stars to be used to do pattern recognition, then there will be $C_3^n = \frac{n!}{(n-3)!3!}$ triplet combination of stars to create triangle patterns. For example if there are 4 stars, namely star 1, star 2, star 3, and star 4, then the triangle patterns is formed by star 1-2-3, 1-2-4, 2-3-4, and 1-3-4. The steps are as follows

a. Choose a closest star to the center of FOV, namely central star.
b. Choose $n-1$ closest stars to the central star, namely star 2, star 3,..., star $n$.
c. Create all possible triplet combination of stars. A simple algorithm to generate all possible triplet combination is as follows

```
for I = 1 : n-2
   for j = i+1 : n-1
      for k = j+1 : n
         combination = [ I j k]
```

```
        end
    end
end
```

For example, if $n = 5$ then the pattern is depicted by Figure 2.

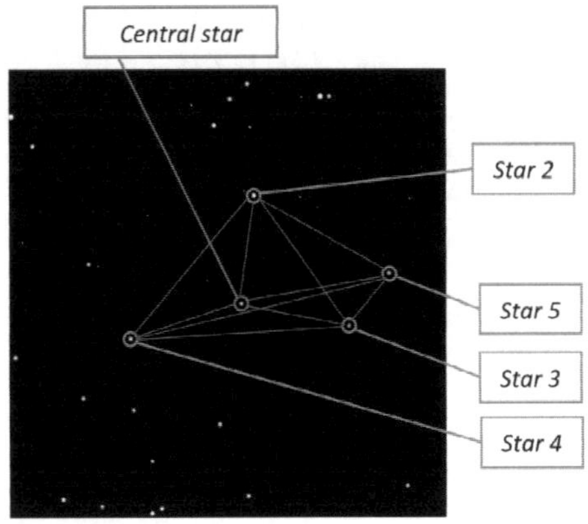


**Figure 2.** Generated pattern for 5 stars in star image.

d. Compute all triplet planar angles in every triangle. Then stored in an array matrix $\boldsymbol{\theta_s}$ which has size $C_3^n \times 3$

$$\boldsymbol{\theta_s} = \begin{bmatrix} \theta_{si11} & \theta_{si12} & \theta_{si13} \\ \theta_{si21} & \theta_{si22} & \theta_{si23} \\ \vdots & \vdots & \vdots \\ \theta_{siC_3^n1} & \theta_{siC_3^n2} & \theta_{siC_3^n3} \end{bmatrix} \quad (3.1)$$

where the values in matrix's row had sorted from the smallest to the largest.

e. For every star $i$ in GSC, choose its $n-1$ neighbour stars. Create all possible triplet combination of stars using step c, and compute all planar angles and stored in an array matrix namely $\boldsymbol{\theta_{ci}}$

$$\boldsymbol{\theta_{ci}} = \begin{bmatrix} \theta_{c11} & \theta_{c12} & \theta_{c13} \\ \theta_{c21} & \theta_{c22} & \theta_{c23} \\ \vdots & \vdots & \vdots \\ \theta_{cC_3^n1} & \theta_{cC_3^n2} & \theta_{cC_3^n3} \end{bmatrix} \quad (3.2)$$

f. Reduce every $\boldsymbol{\theta_s}$'s row with every $\boldsymbol{\theta_{ci}}$'s row and take its total absolute error. If the error is less than a particular value $\varepsilon$, give it a label 1 and if not, then give it a label 0. In this paper we use $\varepsilon = 1°$. The labels are stored in an matrix $\Delta_i$ which has size $1 \times C_3^n$. Do the same step for every star $i$ from star catalogue.

g. Select $\Delta_i$ which has the most number of label 1. The selected $\Delta_i$ is the matched pattern.

In brief, this algorithm works with a voting principle. A pattern which has the most number of matched triangle is the matched pattern. Suppose there are two candidates as shown by Figure 3.

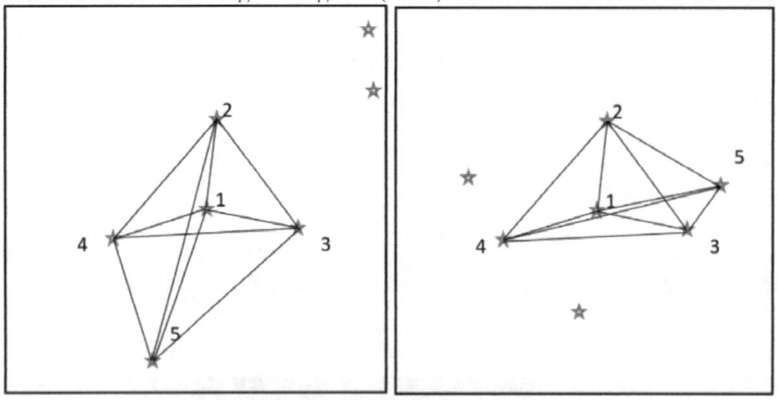

**Figure 3.** Pattern candidate from GSC.

Then the voting scheme is shown as on Figure 4.

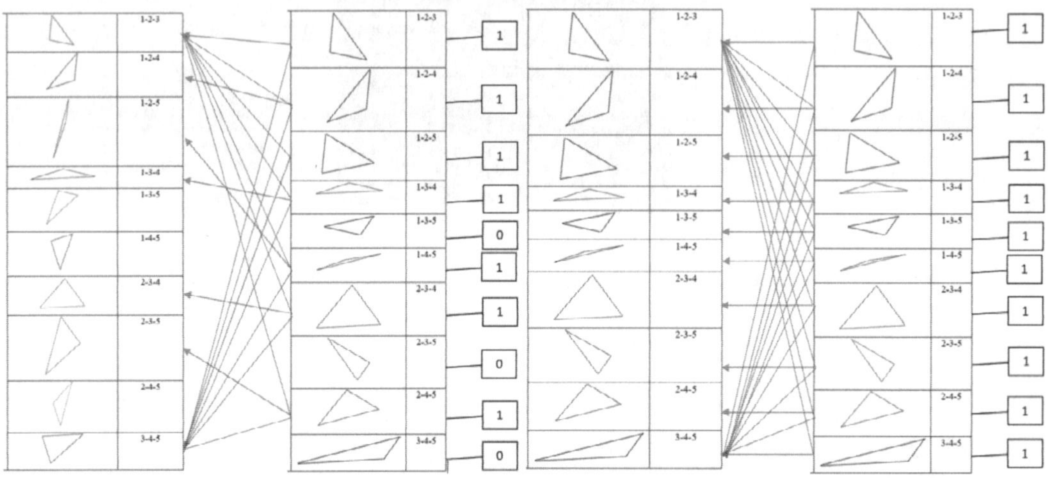

**Figure 4.** Voting scheme between two candidates.

The first column in Fig. 4 is pattern candidate form GSC, while the second column is pattern from star image. The left side is candidate 1 and the right side is candidate 2. The green arrow is for the matched triangle, while red arrow for the unmatched triangle. Candidate 2 has more matched triangle than candidate 2, thus the matched pattern is candidate 2.

This algorithm uses the following two GSC: primary and secondary GSC. Structures are as detailed on Table 1 and Table 2 [10].

**Table 1.** The structure of primary GSC [10]

| ID Star | Right ascension | Declination |
|---|---|---|
| 1 | $\alpha_1$ | $\delta_1$ |
| 2 | $\alpha_2$ | $\delta_2$ |
| ... | ... | ... |
| n | $\alpha_n$ | $\delta_n$ |

**Table 2.** The structure of secondary GSC.

| ID Star | ID neighbour 1 | ID neighbour 2 | ... | ID neighbour m |
|---|---|---|---|---|
| 1 | IDn 1 | IDn 2 | ... | IDn m |
| 2 | IDn 1 | IDn 2 | ... | IDn m |
| ... | ... | ... | ... | ... |
| n | IDn 1 | `IDn 2 | ... | IDn m |

where ID neighbour 1 to ID neighbour m is ordered ascendingly according to its distance from main ID star. We use SAO J2000 Star Catalogue with star whose magnitude is less than 6 as GSC.

## 4. Results and Discussion

Digital simulation are done using the following hardware specifications:

- *Processor*: AMD FX 9830P 3.00 – 3.50 GHz
- RAM: 16 GB
- *Graphic card*: Radeon RX 460 4 GB DDR 4

While the digital simulation setups are as follows:

- *FOV* = 17°
- *Star's magnitude limit in star image = 6*
- *Number of random input boresights = 1000*
- *GSC = SAO J2000*
- *Star's magnitude limit in GSC = 6*

Digital simulations are done in various resolution of star image to see its effect toward success rate, because the lower the resolution, star spot centroiding error will be higher thus the pattern would be mistaken, and vice versa. We varied resolution 256×256, 512×512, 1024×1024, and 2048×2048. The result is shown in Table 3.

**Table 3.** Digital simulation results.

| n stars | Resolution | Success rate (%) | Average Processing time (s) |
|---|---|---|---|
| 4 ( 4 triangles) | 256 × 256 | 58.1 | 1.428472 |
| | 512 × 512 | 84.8 | |
| | 1024 × 1024 | 93.1 | |
| | 2048 × 2048 | 94.8 | |
| 5 (10 triangles) | 256 × 256 | 75.6 | 4.394685 |
| | 512 × 512 | 91.5 | |
| | 1024 × 1024 | 97.8 | |
| | 2048 × 2048 | 98.5 | |
| 6 (20 triangles) | 256 × 256 | 87.8 | 12.62772 |
| | 512 × 512 | 96.8 | |
| | 1024 × 1024 | 98.7 | |
| | 2048 × 2048 | 99.5 | |
| 7 (35 triangles) | 256 × 256 | 92.9 | 30.15096 |
| | 512 × 512 | 99.1 | |
| | 1024 × 1024 | 99.8 | |
| | 2048 × 2048 | 99.9 | |

From Table 3, the algorithm has an acceptable success rate (93.1%) with only 4 stars and 1024×1024 resolution. While in 5 stars and the same resolution, it has success rate 97.8% but the

IOP Conf. Series: Materials Science and Engineering **645** (2019) 012001 doi:10.1088/1757-899X/645/1/012001

processing time is also increased significantly. If we use 6 and 7 stars in the same resolution, the success rate is not increased significantly, but the processing time is increased alot. The interesting part is this algorithm still has high success rate in 6 stars even on low resolution (512×512 with success rate 96.8%) and in 7 stars has 92.9% for 256×256 and 99.1% for 512×512.

The relation between resolution and success rate for every n stars is depicted by Fig. 5.

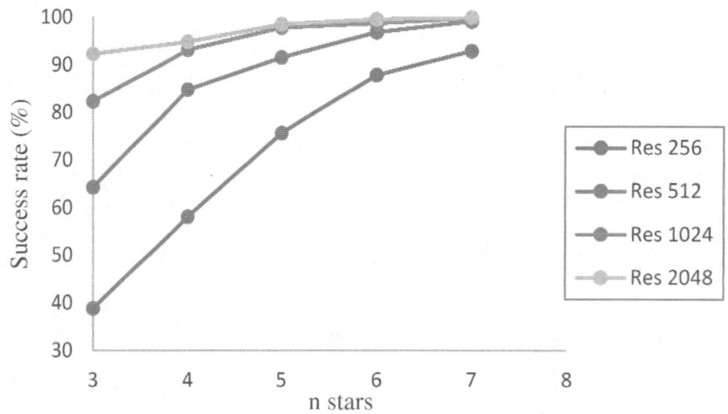


**Figure 5.** Success rate vs n stars for every star image resolution.

The increased processing time as the increased number of star is due to the fact that this algorithm use $C_3^n$ combination triangle patttern. But processing time could be reduced by eliminating step e and store $\boldsymbol{\theta_{ci}}$ in an GSC, thus the calculation of $\boldsymbol{\theta_{ci}}$ is done beyond algorithm.

## 5. Concluding Remarks

From the results and discussion above, this algorithm is capable of doing star pattern recognition and has high level of success rate even on low resolution, Hence this algorithm is expected to still work well on low resolution camera. But the processing time is also increased as the number of star that be used is increased. This increasing processing time could be reduced by calculating $\boldsymbol{\theta_{ci}}$ beyond the algorithm and stored it in a GSC. In the future work, we will implement this algorithm in the hardware and test it using both hardware-in-the-loop simulation and field test of observation.

## References

[1] C. C. Liebe, "Star trackers for attitude determination," *I E E E Aerospace and Electronic Systems Magazine,* 1995.

[2] K. M. HUFFMAN, "Designing Star Trackers to Meet Micro-satellite Requirements," 2006.

[3] S. Dikmen, "Development of Star Tracker Attitude and Position Determination System for Spacecraft Maneuvering and Docking Facility," 2016.

[4] T. Delabie, "Star Tracker Algorithms and a Low-Cost Attitude Determination and Control System for Space Missions," 2016.

[5] R. H. T. M Arif Saefudin, "Algoritma Pengenalan Pola Bintang untuk Deteksi Posisi Bintang pada Star Sensor Satelit LAPAN," *Jurnal Teknologi Dirgantara,* vol. 8, pp. 36-44, 2010.

[6] D. M. F. Landis Markley, "Quaternion Attitude Estimation Using Vector Observations".

[7] J. A. Tappe, "Development of Star Tracker System for Accurate Estimation of Spacecraft Attitude," 2009.

[8] J. L. C. Craig L. Cole, "Fast Star-Pattern Recognition Using Planar Triangles," *Journal of Gudance, Control, and Dynamics,* 2006.

[9] G. Zhang, Star Identification Methods, Techniques, and Algorithms, Berlin: Springer, 2017.

[10] M. A. Saifudin, "Pembuatan Katalog Bintang untuk Aplikasi Star Sensor menggunakan Metode Magnitude Filtering".

IOP Conf. Series: Materials Science and Engineering **645** (2019) 012002 doi:10.1088/1757-899X/645/1/012002

# Texture Traits with Uniform-Quantization in Handwriting Documents for Digital Forensics Investigation

**D Pratiwi[1,*], Syaifudin[1], T Rahardiansyah[2], A Hilman[1], W Anggriani[1], N Chairunnisa[1]**

[1] Department of Informatics Engineering, Trisakti University, Indonesia
[2] Department of Law, Trisakti University, Indonesia

[*] Email: dian.pratiwi@trisakti.ac.id

**Abstract**. In crime and falsification related to documents are often difficult to verify orauthenticated by the authorities. This is the basis of research to develop a system in assisting digital forensics to investigate and seek the truth of the evidence in the form of digital handwritten documents. The steps that being taken by researchers are collecting and digitizing the documents into image form, converting color from RGB to greyscale, separate the object through thresholding, color histogram, uniform quantization from 256 to 128 of greylevel, texture feature extraction ie variance, skew, relative smoothness, entropy and mean, normalize the feature value and similarity measures through Euclidean distance calculations. From the results of testing of 10 data that has been matched with 20 training data, 6 documents successfully recognized correctly the authenticity of the document owner. Thus, the system in this study produces an accuracy of 60% and can be used to assist digital forensics in analyzing the authenticity of handwritten documents.

## 1. Introduction

Currently, technology has an important role of various area of life. Technology can be used not only for connect people and build relationship around the world, but also to facilitate in completing a job or various problems. For example in public service area, technology has been using in Akshaya Program [2] to obtain a driver's license or keep the government accountable. Other field, for example, in the healthcare industry, medical technology has been used by many healthcare practitioners to improve their practice from better diagnosis, surgical procedures and improved patient care. And in this study, researcher focus on develop some kind technology in law area, especially for digital forensics to help them solve the problems and various crime cases, especially those involving digital handwritten document fraud crimes in Indonesia.

A Research which related to handwritten document, ever done also by other researchers. As in the study of Plamondon and Srihari [5], which developed a technique of recognition of every word in handwriting either from online or off line through the Hidden Markov method with good results. Then in Gluchev, G [6] various methods of feature extraction approach were introduced to recognize handwriting features as a way of investigating the authenticity of the owner's handwriting. Such as predominant slope feature, quantity of movement, direction of movement, and many others. Research on handwriting analysis has also been done by researchers in 2007 [7] where the researcher succeeded in processing handwritten documents to know the personality type of the owner through RegionbasedShape Feature Extraction with accuracy reaching 81.6%. Then in 2016 [8], researchers have also succeeded in developing a prototype of handwritten document security through Inner Product method with accuracy result obtained from 100 testing document by 72%. From this result, the researcher tries to develop again with the same research theme, where the approach is to use Statistical TextureFeature Extraction combined with Uniform Quantization, and similarity measure technique through Euclidean Distance. Selection of texture-based feature extraction method is due to Chandy's research [9], the result of accuracy is very good that is 85.1% when using texture as input value of irregular, asymmetric, and ill-defined data. Just like everybody's handwriting, which has a characteristic difference in every stroke. This is what then researchers try to apply to handwritten document to produce unique features so that it can distinguish well the handwriting owner to one another. These features are the fineness of strokes, variations of writing, slope and irregularity strokes.

## 2. Materials and Methods

In this research, we used several methods to develop a mobile system that can match the needs of investigations in the field of digital forensics.

### *2.1. RGB Color Convertion*

RGB color convertion in this study is the stage used to change the true color (24-bits) in handwriting form into greyscale color (8-bits), by calculating the mean value of *R (Red), G (Green),* and *B (Blue)* at each pixels so that it only has a colour interval between 0 and 255 [7].

### *2.2. Thresholding*

Thresholding is one of segmentation methods in image processing which used to group pixels within certain intensity limits. In addition, thresholding also serves to separate the image that matches the object (foreground) and background, and convert the image into binary image to facilitate the next process. In this study, thresholding is used to separate the written object from the background, and the selected value is the middle value of color interval, which the value is 128.

### *2.3. Color Histogram*

A histogram is a graphical representation that shows the gray color distribution in an image [10]. The histogram is calculated by the formula :

$$h_i = \frac{n_i}{n}$$

where $n_i$is the number pixel that have a gray value $i$ ($i = 0 \ldots L-1$), $L$ is maximal color interval, $n$ is amount of pixels in image and$h_i$ is a probability of $i$ gray value [11]. In this research, grey level which used will begrouped again into a certain color scale through the quantization stage.

### *2.4. Quantization*

Quantization is a part of the image digitization stage that classifies the grey level value of continuous image to a certain level. Quantization can divide the large quantity into a discrete number of small parts, often assumed to be integral multiple of a common quantity. There are two types of quantizer, *Non-uniform Quantizer* and *Uniform Quantizer* [12]. Uniform quantizer has the same gray level grouping interval (for example, intensity from 1 to 10 is rated 1, intensity from 11 to 20 is rated 2, and so on). While the non-uniform quantizer is a finer quantization required especially in the image portion that describes the detail or texture or boundary of an object region, and the more rough quantization is applied to the same region on the part of the object. For example, on a satellite photograph of the earth's surface, where densely populated areas are finely quantized, while sea areas are roughly quantized. In this research, the quantization technique which used is *Uniform Quantizer* because handwriting data tend to have simple pattern and detail (not many objects). The formula is $G=2^m$, where $G$ is grey level and $m$ is number of bit.

Quantization will determine the brilliance resolution of an image. If the scale used is too small, the image resolution will be smaller and it may cause blurry, broken or unclear to the image.

### *2.5. Feature Extraction*

Feature extraction (or sometime called by indexing) is a basic step in conducting an image interpretation and classification. There are many ways to extract the feature, depends on the data. In image data, features that can be extracted are color, shape, and texture [17]. And in this study, researcher uses texture. Texture is an intrinsic character of image which related with roughness level, granulation, and regularity of structural arrangement of pixels. Texture is defined as spatial distribution of greylevel in a set of neighbouring pixels. Image feature extraction based on texture in orde one can use statistical method, ie by looking at the greylevel distribution statistic on the image histogram [13]. From the histogram values, it can calculate feature parameters :

a) Variance ($\sigma^2$)

Indicates element variations on the histogram of an image. Variance value is used by researchers to assess the extent of word variaton which present in each handwritten document. The greater the value of variance, the more varied the existing pattern in writing. The formula is :

$$\sigma^2 = \sum_n (f_n - \mu)^2 p(f_n)$$

Where is an average value of pixels on image, is a value of grey intensity, $(p(f_n)$ is a value of histogram (intensity occurrence probability in document image) and is $\sigma^2$a value of variance

b) Skewness ($\alpha_3$)

Skewness will indicate the relative historic level of the histogram curve of an image. In this research, value of skewness will be used to assess the inclination of the stroke direction of a post on each document. If the value is close to 0, the direction of the skew is symmetric. The formula is:

$$\alpha_3 = \frac{1}{\sigma^3} \sum_n (f_n - \mu)^3 p(f_n)$$

c) Entropy (*H*)

Entropy will indicate irregularity level of a pattern, and this is also will be used in research on handwritten document. The higher value of entropy, the variations and information contained in the writing pattern more and more irregularity.

$$H = -\sum_n p(f_n). Log_2 p(f_n)$$

d) Relative Smoothness (*R*) [14]

Relative smoothness is a value that will indicate the relative degree of smoothness of shape and pattern of an image, and the researcher would also use it to determine how fine or rough the handwriting strokes of each document are. The higher the value of relative smoothness, the smoother the strokes pattern and the faster the writing movement.

$$R = 1 - \frac{1}{(1 + \sigma^2)}$$

In this research, another feature added is the mean feature, which will indicate the exact feature value to represent the entire representation of handwriting patterns.

### *2.6. Normalization*

Normalization is a step to create a range that has different value to the same scale and smaller as well. There are many ways to normalize value. One of the best method is Min-Max Normalization [16] :

$$D`(t) = \frac{D(t) - \min(D) \cdot (U - L) + L}{\max(D) - \min(D)}$$

where D is the natural data, U and L are the upper and lower normalization bound.

### *2.7. Similarity Measure*

Similarity measures are techniques that used to measure the level of similarity between objects to one another, and usually implemented on pattern recognition. This similarity level is determined by distance. From the scientific and mathematical point of view, distance is defined as a quantitative degree of how far apart two objects are [15]. There are several similarity measures techniques, ie *Eulidean, City Block, Minkowski, Chebyshev, Sorensen, Gower, Soergel, Kulezynsky, Canberra*, andothers. In this study, the technique chosen is *Euclidean* because *Euclidean Distance* is the only metric that is the same in all direction. Its fits very nicely with the general qualities of our universe, which is also rotation invariant.

$$d_{Euc} = \sqrt{\sum_{i=1}^{d} |P_i - Q_i|^2}$$

### *2.8. Research Methods*

First of all, researcher collects the data in the form of handwritten document (printed) from several source, where the writing should be written on a white background paper without any lines or ornaments). All of documents which collected amounted to 30 from 10 different authors. From 30 data will be divided into 2 categories, namely 20 training data and 10 testing data. All of data is then will be processed into digital form through scanner, and saved as .jpg with the size of 200x150 pixels. After that, the data will be processed into pre-processing step, started from converting RGB color to greyscale, and then separate the writing area from the background through thresholding techniques. The given threshold value is 120, where this value is taken based on trial and error testing of the best results through the researcher's vision (subjective). Then, pixel's value of each document, will be counted by the color probability of histogram, ranging from grey level 0 to 255. After that, grey level will be quantized into $2^7$ or 128 (0 to 127) where every 2 levels of gray will be represented by 1 level. This is to be done to reduce the computational load in finding the texture value at the feature extraction stage. Then from the result of quantization will be proceed to the texture feature extraction to get the value of *variance, skewness, entropy,* and *relative smoothness*. After that, *mean* value will also be used as an additional feature.

This process is applied to all handwriting data, both training data and testing data (digital handwriting query), which is then matched through the similarity measure technique after normalization into [0.0 – 1.0] interval. The smallest distance obtained from the results of this calculation will determine the similarity of documents between the authors one to another. The results from this stage then will be calculated the percentage of accuracy to determine how well the system developed by researchers.

## 3. Results and Discussion

The system which developed in this research is using Android Studio based on Java programming language. Figure 1 is the view of system. All training data is processed to take its features through the system, and stored in SQLite database which will be used as a reference to determine the owner of document against testing data. The difference in feature intervals will be automatically changed by the system into 0 to 1 value. If there is a feature whose value cannot be defined or NaN (as shown in Figure 1b), the default value of the feature is 0. Table 1 shows the result of measuring the distance between testing and training data using euclidean distance.

IOP Conf. Series: Materials Science and Engineering **645** (2019) 012002 doi:10.1088/1757-899X/645/1/012002

(a) (b) (c)

**Figure 1.** Mobile System (a) Main Menu; (b) Register Menu; (c) Checker Menu

**Table 1.** Similarity Measure of Handwritten Documents.

| No | Distance Value | Target | | | | | | | | | |
|---|---|---|---|---|---|---|---|---|---|---|---|
| | | B | B | D | D | H | H | I | I | J | J |
| 1 | A | 0.73 | 0.75 | 3.41 | 3.59 | 3.32 | 3.2 | 4.52 | 2.22 | 3.21 | 3.18 |
| 2 | A | 1.28 | 1.19 | 3.28 | 3.05 | 3.39 | 3.34 | 4.59 | 2.33 | 3.25 | 3.16 |
| 3 | B | 0.33 | 0.35 | 2.44 | 2.33 | 2.92 | 2.71 | 3.09 | 1.38 | 2.51 | 2.4 |
| 4 | B | 0.37 | 0.32 | 2.46 | 2.4 | 2.97 | 2.6 | 3.07 | 1.41 | 2.42 | 2.37 |
| 5 | C | 4.12 | 4.08 | 2.02 | 2.1 | 2.41 | 2.31 | 1.03 | 3.1 | 2.1 | 2.07 |
| 6 | C | 4.05 | 4.11 | 2.08 | 2.09 | 2.43 | 2.32 | 1.2 | 1.2 | 2.12 | 1.98 |
| 7 | D | 1.75 | 1.8 | 0.51 | 0.01 | 0.8 | 0.10 | 1.18 | 1.3 | 0.41 | 0.35 |
| 8 | D | 1.46 | 1.76 | 0.98 | 0.88 | 1.12 | 1.6 | 2.6 | 0.4 | 1.41 | 1.23 |
| 9 | E | 3.45 | 3.3 | 1.12 | 1.17 | 1.14 | 1.62 | 0.72 | 2.4 | 1.31 | 1.3 |
| 10 | E | 0.86 | 0.78 | 1.15 | 1.01 | 1.80 | 0.98 | 2.01 | 0.31 | 1.4 | 1.35 |
| 11 | F | 0.68 | 0.77 | 3.02 | 3.6 | 1.89 | 3.51 | 4.4 | 2.33 | 3.24 | 3.07 |
| 12 | F | 1.19 | 1.20 | 1.02 | 1.14 | 1.4 | 1.3 | 2.4 | 0.14 | 1.03 | 1.04 |
| 13 | G | 3.45 | 3.42 | 1.05 | 1.71 | 1.42 | 1.6 | 0.69 | 2.4 | 1.3 | 1.34 |
| 14 | G | 1.03 | 1.06 | 1.23 | 1.08 | 1.4 | 1.02 | 2.08 | 0.33 | 1.44 | 1.27 |
| 15 | H | 2.07 | 2.11 | 0.41 | 0.31 | 0.5 | 0.21 | 1.3 | 1.04 | 0.15 | 0.23 |
| 16 | H | 4.27 | 4.22 | 2.23 | 2.51 | 2.3 | 2.47 | 1.5 | 3.12 | 2.13 | 2.01 |
| 17 | I | 1.42 | 1.45 | 3.1 | 3.7 | 3.1 | 3.62 | 4.7 | 2.43 | 3.3 | 3.2 |
| 18 | I | 2.31 | 2.37 | 1.4 | 1.11 | 1.8 | 1.2 | 0.09 | 2.4 | 1.51 | 1.34 |
| 19 | J | 1.02 | 1.05 | 1.12 | 1.31 | 1.5 | 1.22 | 2.3 | 0.03 | 1.13 | 0.14 |
| 20 | J | 1.52 | 1.66 | 1.09 | 1.82 | 1.01 | 1.79 | 2.81 | 0.54 | 1.41 | 0.09 |

***Note*** *:A: Agung, B : Anung, C: Binti, D: Dian, E: Is, F: Najih, G: Ratna, H: Sabil, I: Syaifudin, J: Syaikhu*

Based on the test result in table 1 above, the error of handwritten document recognition still occur, for example in experiments number 7, 15, and 19. This can be due to the features extracted between documents D with H, document J with H, document I with J having no distant value difference. There was a written similarities in terms of speed, direction of stroke, irregularities and variations. But in term of accuracy, this study is still better at recognition handwritten patterns when compared with previous studies. In the previous study [8], the accuracy achieved was only 55% (11 out of 20 data), whereas in this study, 6 out of 10 testing documents were identified correctly or achieved accuracy of 60%. This accuracy may increase if the data used as reference (in SQLite) more, because the value of features to be matched will be more varied to allow details of the different features of the posts will be more visible and the system can more easly recognize the pattern. So that, the system built by researchers can be utilized for the purposes of digital forensic in helping investigate the evidence in the form of authenticity of handwritten document ownnership with a fairly good level of accuracy.

## 4. Conclusion

Implementation of feature extraction based on texture with uniform quantization in the system development for digital forensics has a success rate that is good enough for 60%, to prevent the forgery of handwritten documents and investigate the origins of writing, because of the testing results of 10 documents written from 5 different authors, 6 documents were correctly identified by the authenticity of the owner. From the five texture features which have been used to investigate the authenticity of handwritten, *the relative smoothness* is the most difficult feature applied to differentiate document ownership because it has the same value results for all tested data. This can be due to *the relative smoothness* value associated with flexibility or speed, where handwriting is essentially created automatically by individual capabilities or without fabrications, so that all documents have maximum relative smoothness value of 1. Texture features that can play a good role in distinguishing characteristics of handwriting features are *skewness, entropy*, and *variance*. While the mean value may change depending on the number of words contained in the document and the size of writing.

## Acknowledgments

This research was fully funded by Ministry of Technology Research DIKTI, Republic of Indonesia, as a Grant for lecturer, and also highly supported by Trisakti University and dedicated to author's beloved parents, especially mrs Sri Mulyani (deceased).

## References

[1] Marquez, S. et al. 2008. A Simple and Effective Method of Color Image Quantization. *LectureNotes in Computer Science (LNCS)*, Vol.5197, pp. 749-757.

[2] Kumar, R. Five ways technology is improving public service. 2014. *The World Bank blogs.* [Online]. Available: http://blogs.worldbank.org/governance/five-ways-technology-improving-public-services.

[3] Ramadhani, Y. 2018. Lion Air Laporkan 9 Pilot dan 1 Karyawan Atas Pemalsuan Dokumen. Tito.id. Available: https://tirto.id/lion-air-laporkan-9-pilot-dan-1-karyawan-atas-pemalsuan-dokumen-cKWS.

[4] Santoso, A. 2015. Perjalanan Kasus Kematian Akseyna, Setahun Berlalu Masih Misteri. *Liputan6.* [Online]. Available:http://news.liputan6.com/read/2472824/perjalanan-kasus-kematian-akseyna-setahun-berlalu-masih-misteri.

[5] Plamondon, R. and Srihari, S.N. 2000. On-Line and Off-Line Handwriting Recognition : A Comprehensive Survey. *IEEE Transactions on Pattern Analysis and Machine Intelligence*, Vol.22, No.1, pp. 63-84.

[6] Gluhchev, G. 2004. Handwriting in Forensic Investigations. *International Journal "InformationTheories & Applications",* Vol.11, pp. 42-46.

[7] Pratiwi, D., Santoso, G.B., Saputri, F.H. 2017. The Application of Graphology and Enneagram Techniques in Determining Personality Type Based on Handwriting Features. *Journal ofComputer Sciences and Information*, Vol.10, No.1.

[8] Syaifudin and Pratiwi, D. 2016. Security Handwritten Documents by Using Inner Product. *Lecture Notes in Electrical Engineering* 365, Springer Science+Bussiness Media Singapore,DOI 10.1007/978-981-287-988-2_56.

[9] Chandy, D.A., Johnson, J.S., Selvan, S.E. 2014. Texture Feature Extraction using Gray Level Statistical Matrix for Content-based Mammogram Retrieval. *Springer Sciences+BusinessMedia New York : Multimedia Tools and Application*, Vol. 72, Issue 2, pp. 2011-2024.

[10] Purnomo, A., Puspitodjati, S. 2010. Aplikasi Pemrograman C# untuk Analisis Tekstur Kayu Parquet dengan Menggunakan Metode Grey Level Co-occurrence Matrix (GLCM), [Online]. http://www.gunadarma.ac.id/library/articles/ graduate/industrial-technology/2009/Artikel_50405312.pdf.

[11] Najeeb, H. 2017. Histogram of Color Image. Mathworks : Matlab Answers. [Online]. Available:https://www.mathworks.com/matlabcentral/answers/324646-histogram-of-color-image.

[12] Shodhganga. 2017. Quantization Techniques. [Online]. Available: http://shodhganga.inflibnet.ac.in/bitstream/10603/25341/8/08_chapter%203.pdf.

[13] Wong, J.S.J., Zrimec, T. 2006. Classification of Lung Disease Pattern Using Seeded Region Growing, *Lecture Notes in Artificial Intelligence : Proceedings 19$^{th}$Australian Joint Conferenceon Artificial Intelligence*, pp.233-242, Berlin, Germany : Springer-Verlag.

[14] Singh, B.Kr., Mazumdar, B. 2010. Content Retrieval from X-ray Images Using Color & Texture Features, *International Journal of Electronics Engineering*, 2(1), 25-28.

[15] Cha, S.H. 2007. Comprehensive Survey on Distance/Similariy Measures between Probability Density Functions. *International Journal of Mathematical Models and Methods in AppliedSciences*, Issue 4, Vol. 1, pp. 300-307.

[16] Pratiwi, D., Santika, D.D., and Pardamean, B. 2011. An Application of Backpropagation Artificial Neural Network Method for Measuring The Severity of Osteoarthritis. *InternationalJournal of Engineering & Technology IJET-IJENS*, Vol.11, No:03, ISSN : 117303-8585.

[17] Davin, R.P., Pratiwi, D., Syaifudin, Trubus. R., Rizky, D.L.P., 2017. Implementation of Inner Product to Analyze Digital Handwriting based on Texture Traits. Proceedings of The 2017 International Conference on Computer Science and Artificial Intelligence (CSAI), pp. 114-118, ACM, ISBN: 978-1-4503-5392-2.

IOP Conf. Series: Materials Science and Engineering **645** (2019) 012003 doi:10.1088/1757-899X/645/1/012003

# Development of Hybrid Simulation Model of Air Traffic Management in the Terminal Control Area

**R Medianto**[1], **H M Pasaribu**[1,*], **and H Muhammad**[2]

[1]Aircraft Design, Operation, and Maintenance Research Group, Faculty of Mechanical and Aerospace Engineering, Bandung Insitute of Technology, Bandung, Indonesia

[2]Flight Physics Research Group, Faculty of Mechanical and Aerospace Engineering, Bandung Insitute of Technology, Bandung , Indoneisa

*E-mail: hisar.m.pasaribu@gmail.com

**Abstract.** The air traffic density at the Terminal Control Area can lead to several problems related to the safety and efficiency of flight operation if not adequately managed. These problems include the increased risk of crashes between aircraft, airborne and ground delays, waste of fuel and environmental problems caused by gas emission and noise. One of the analysis tools to solve these problems is the hybrid simulation model. The socio technic characteristics of the system can be observed more intact using a hybrid simulation model so that the results of the analysis more precise and comprehensive. This article describes the research framework and progress on the development of the hybrid simulation model of an ATM system in the Terminal Control Area. The final model will integrate of agent-based simulation, discrete event simulation, and system dynamics models. An agent-based simulation model that defines the movement of aircraft in the Terminal Control Area has been developed in this early research phase. The initial model can describe the rules for minimum separation between aircraft that are applied to ensure the safety of flight operations. The following aircraft will reduce its speed so that an adequate separation distance is obtained with the aircraft in front of it. The model then will be improved by integrating a discrete event simulation and system dynamics models on the initial model to obtain a complete hybrid simulation model.

## 1. Introduction

Passenger of world air transport has doubled every 15 years since 1977, and according to the International Civil Aviation Organization (ICAO), this growth continues [1]. Especially in the Asia Pacific region, the growth of air transportation in the past decade shows a high rate which is estimated to reach 37% of the world's air transportation market in 2017 from 33% in 2012 [2]. This is also supported by an increase in the addition of aircraft in the Asia Pacific region which will reach 37% in 2036 with the number of jet aircraft reaching 17,520 from the previous 6,830 in 2016 [3].

On the other hand, the growth brings its challenges, mainly resulting in air traffic congestion in the Terminal Control Area of several major hub airports. The higher density of flight traffic increases the risk of collisions between aircraft. Soekarno-Hatta International Airport (CGK), which is one of the busiest airports in the Asia Pacific, has experienced an increased risk of collisions and delays. This was conveyed by representatives of the Indonesian Air Traffic

Controllers Association (IATCA) [4] which stated that an increase in aircraft movements at CGK Terminals Control Area that were not matched by infrastructure availability resulted in prolonged delays and endangered flight safety.

The Terminal Control Area provides control on airspace around a large/busy airport; it has the highest level of Air Traffic Services including conflict detection services, separation determination, sequencing, and spacing. In the Terminal Control Area runs three flight operation scenarios at the same time they are arrival, departure and crossing flight [5]. The density of flight traffic at the Terminal Control Area can cause some problems in flight safety and operating efficiency if not properly controlled. These problems include increased risk of aircraft collisions, delays, waste of fuel and environmental disturbances caused by emissions and noise pollution.

In this research, a hybrid simulation model will be developed by integrating agent-based simulation, discrete event simulation, and system dynamics simulation models. Hybrid simulation modeling can better represent the characteristics of a complex system such as the ATM system [6]. The development of computational technology has also helped the reliability of this method in handling complex calculations which are a decrease of the complex phenomena of the modeled system. The hybrid simulation model will then be used to analyze the risk of collisions at CGK Terminal Control Area.

## 2. Research Methodology

### *2.1 ATM Hybrid Simulation*

ATM systems have a large scale of problems as there are several human operators in the system (pilots, controller, and others) who each have the freedom to make decisions. By combining two simulation modeling methods, it is expected to reduce the limitations and improve the ability of the simulation method. Combining these two simulation coding methods is called hybrid simulation modeling [7].

The application of hybrid simulation modeling in the ATM research field is relatively recent. Wei and Li (2016) analyzed ATM systems from aspects of the internal operating mechanism and external behavior attributes [6]. In this research, the ATM hybrid simulation model was developed which was represented by the integration of DES and ABS. The ABS model is used to describe the micro-behavior of individual flight traffic and integrate the micro-behavior into dynamic macro performance at the system level. Agents included in the ABS simulation model include aircraft and flight traffic control personnel. The hybrid model is simulated on arrival and departure operations. Validation results show that the ATM hybrid simulation model has high accuracy regarding micro behavior and macro system performance.

The integration of DES models and SD models was carried out by Glover and Lygeros in 2004 in the form of an ATC model for the certainty of three-dimensional separation [8]. In this model, the aircraft is represented by a model of flight dynamics in the form of an airplane system dynamics model. DES model represents flight plans and logic variables in the Flight Management System (FMS). Then in 2016, Pasaoglu et al. developed the model by including the ATC model that was built with the paradigm of the DES model. The algorithm built has polynomial complexity related to the number of aircraft and waypoints [9].

### *2.2 System Description*

The ATM system in the Terminal Control Area modeled in this research is limited to arrival operations only. The ideal arrival pattern of two-dimensional views in a Terminal Control Area serving two parallel runways is shown in Figure 1. The aircraft will enter the Terminal Control Area through one of four entry points, namely E1, E2, E3 or E4. The entry point is the transition points between en-route airspace and Terminal Control Area.

The determination of entry points coordinates in conventional systems is assisted by a system of land-based navigation aids such as Non-Directional Beacon (NDB), VHF Omni-directional Range (VOR) and Distance Measurement Equipment (DME). In a more advanced system using the Performance Based Navigation concept with R-NAV (Area Navigation) and RNP (Required Navigation Performance) systems, the positioning of these points is also assisted by satellite-based navigation systems such as the Global Positioning System (GPS). At these checkpoints, it is also possible for aircraft to hold while waiting for its turn to enter the Terminal Control Area.

Aircraft entering the Terminal Control Area then fly towards runway by following certain trajectories defined with the help of navigation equipment, either land-based navigation equipment (NDB, VOR / DME) or satellite (GPS). The path is often referred to as Standard Terminal Arrival Routes (STAR). With guidance from Flight Management Computer (FMC) aircraft can easily fly along the track. In certain conditions, if needed, the aircraft ATC directives that contain the aircraft heading, height and also changes in speed or often referred to as vectoring.

Aircraft movements at Terminal Control Area at airports equipped with surveillance facilities such as Radio Detecting and Ranging (RADAR) will always be monitored by an air traffic controller. Through the assistance of RADAR, an air traffic controller will always keep the separation between aircraft to be maintained by the applicable separation procedures. For the arriving aircraft, the most commonly used separation is horizontal separation. While between departing and arriving aircraft usually vertical separation is utilized [10].

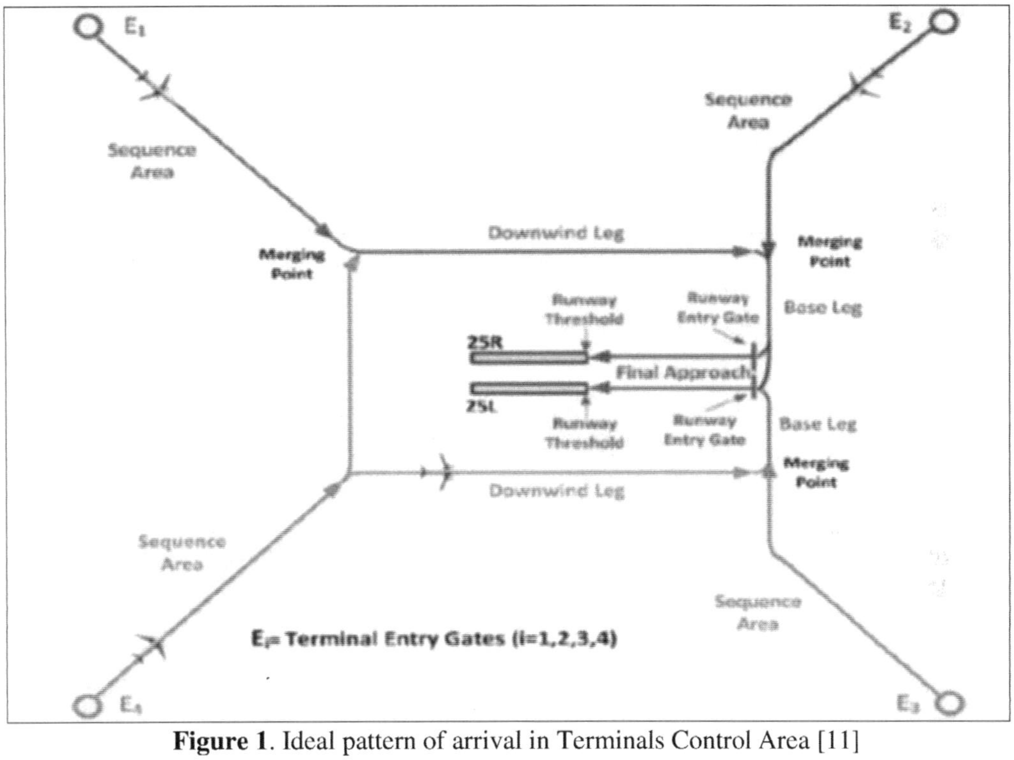


**Figure 1**. Ideal pattern of arrival in Terminals Control Area [11]

Terminal Control Area also applies speed restrictions to certain segments. The closer the runway is, the lower the speed limit. The trajectory segments will likely also meet at certain points which are often referred to as merging points. Terminal Control Area path will end at the runway's entry gate where the final approach path begins. The final approach path and the runway generally are integrated into the runway system.

### 2.3 *Modeling Concept*

ATM system simulation modeling will apply the concept of a hybrid simulation model. The modeling concept that integrates ABS, DES and SD models is outlined in Figure 2.

IOP Conf. Series: Materials Science and Engineering **645** (2019) 012003 doi:10.1088/1757-899X/645/1/012003

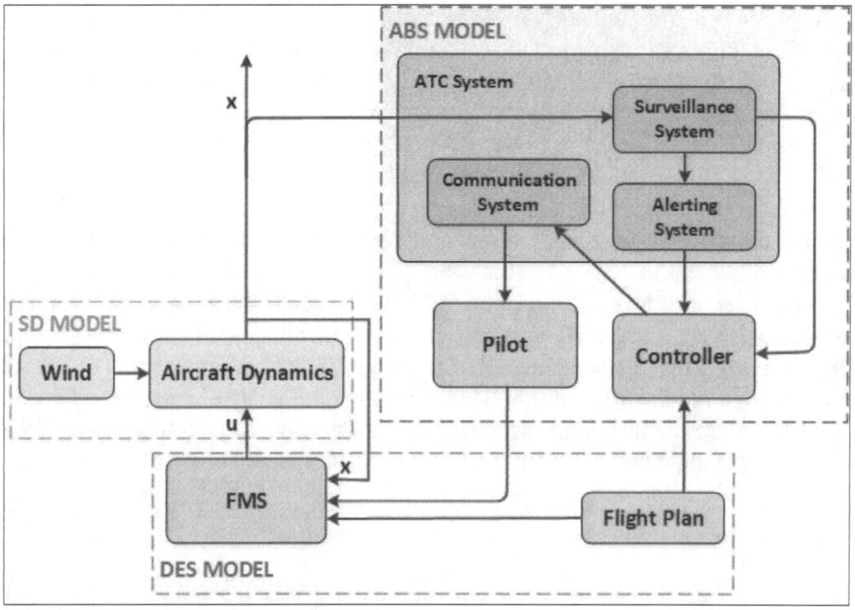


**Figure 2.** Modeling concept

*2.3.1. Agent-Based Simulation Model (ABS).* The ABS model is used to model the dynamics of decision making and interaction between agents who are representatives of air traffic controller, pilot and ATC system. Each agent is given the ability to make decisions and take action under different conditions to achieve its goals. Each agent also can interact with other agents. The human agent model (controller and pilot) used in the research is a human agent model developed by Bouarfa et al. (2013). The model has three main entities, consist of situation awareness, information processing, and cognitive control.

Situation awareness describes agent perceptions of influential elements, understanding of current conditions and future status projections [12]. Errors can occur in processing information because of the limited processing capacity of a human. Situational awareness and workload will affect the cognitive control mode, whether strategic, tactical, opportunistic or scrambled [13].

The ATC system model is a non-human agent that has a surveillance system, alerting system and communication system models [13]. The surveillance system model provides information on the estimated position and speed of the aircraft. This information is then used by the controller to guide traffic, especially to keep a safe separation between aircraft. The system will elicit an alerting if a critical situation occurs in the form of a reduced separation distance that is considered dangerous based on information from the monitoring system. The communication system model bridges the communication between the pilot and the air traffic controller. The ATC system has a system condition parameter in the form of varying levels of reliability and availability.

*2.3.2. Discrete Event Simulation Model (DES).* Flight plan modeling and Flight Management System (FMS) are represented using the DES model. Flight plan will determine the aircraft's arrival time at a particular waypoint whereas the FMS functions as an aircraft control system responsible for generating inputs (u) based on the state variables (x), flight plan information and the action of the pilot who gets orders from ATC [9].

There are four discrete state variables in the FMS model, namely flight level (FL), way-point index (WP), acceleration mode (AM) and climb mode (CM). FL represents a renewed flight altitude based on $x_3$ conditions. FL is also used to determine the nominal speed ($V_{nom}$). WP represents the number of waypoints in the flight plan. AM shows whether the plane is increasing or reducing speed, or is flying at a constant speed. CM determines whether the aircraft is climbing or descending or leveling.

*2.3.3. System Dynamics Simulation Model (SD).* System dynamics simulation models are used to model aircraft dynamics. The aircraft model to be developed has three control inputs which consist of engine thrust ($u_1$), turn angle ($u_2$) and flight path angle ($u_3$). Five state variables that build airplane models consisting of horizontal positions ($x_1$ and $x_2$), altitude ($x_3$), true airspeed ($x_4$), heading angles ($x_5$), and aircraft mass ($x_6$). Wind velocity is also modeled, $W = (w_1, w_2, w_3)$ which is a representation of disturbances from the environment [8].

$$\dot{x}_1 = x_4 \cos(x_5) \cos(u_3) + w_1 \quad (1)$$

$$\dot{x}_2 = x_4 \sin(x_5) \cos(u_3) + w_2 \quad (2)$$

$$\dot{x}_3 = x_4 \sin(u_3) + w_3 \quad (3)$$

$$\dot{x}_4 = -\frac{C_D S \rho(x_3) x_4^2}{2x_6} - g \sin(u_3) + \frac{u_1}{x_6} \quad (4)$$

$$\dot{x}_5 = -\frac{C_L S \rho(x_3) x_4}{2x_6} \sin(u_2) \quad (5)$$

$$\dot{x}_6 = -\eta u_1 \quad (6)$$

$C_L$ and $C_D$ are lift and drag coefficient, S total surface area of the wing, $\rho$ ($x_3$) air density and $\eta$ is the thrust coefficient relative to fuel consumption.

*2.3.4. Hybrid Simulation Model.* After the three simulation models above have been successfully built, the three are integrated into a hybrid simulation model that can better represent the ATM system in the Terminal Control Area. The hybrid simulation model will be able to describe the air traffic dynamics in the ATM system that is represented by separation control between aircraft and aircraft dynamics controlled by FMS. Human factors are represented by the dynamics of human interaction (pilots and controllers) and humans with tools (pilots, controllers and ATC systems).

## 3. Results and Discussion

At the stage of developing the ABS model, each flow of aircraft arrival can be divided into several segments with a certain maximum speed. Each segment connects two waypoints. Each waypoint is a different agent. An aircraft that is also an agent will experience heading changes and speed when reaching a specific waypoint according to the RNAV STAR procedure that applies in that segment.

The system modeled in this study is the arrival operation at Soekarno-Hatta Airport. Development of ABS models using NetLogo. The elements in the model are defined as agents consisting of aircraft and waypoints. The aircraft has properties in the form of initial position, agent form, differentiating color, heading, and speed while waypoint has properties in the form of initial position, distinguishing color, shape, and position.

The aircraft moves to follow the STAR route through predetermined waypoints at a certain speed by the applicable speed limit. The aircraft is ordered to change the heading towards the next waypoint when it has been through a waypoint. The aircraft is also programmed in such a way that it can be precisely at the coordinates of a waypoint before changing the heading towards another waypoint. Also, the aircraft can maintain its separation from other aircraft by slowing or increasing its speed. When the simulation starts, the button on the interface will record the agent property in graphical form to determine the changes that occur throughout the simulation. The display of the simulation model is shown in Figure 3.

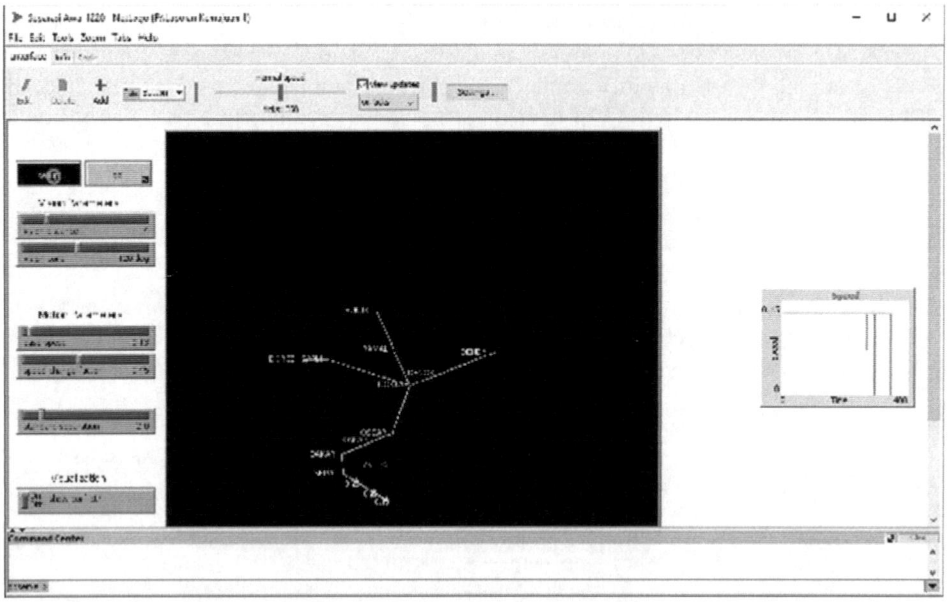

**Figure 3.** Display of Simulation Model of ATM System in Terminal Area

Longitudinal separation applies to the same track, reciprocal tracks and crossing tracks with separate provisions of 5 NM. The aircraft can detect other aircraft in front of it with a radius of 90° to the left and the right. The follower will reduce the speed with certain change factors if there is a conflict (the distance with the preceding aircraft is less than 5 NM) and a potential conflict warning sign will appear (red color).

The simulation model still has several limitations including the separation regulation strategy is only by making a speed reduction. The arrival of the aircraft can only be done at the same time; it cannot be arranged according to a specific schedule. The speed of the plane on a fixed route will only change if there is a potential conflict. Reduction of aircraft speed has not been made gradually (gradually).

The simulation model has been able to run well, even though it is still necessary to validate the model whether it has resembled the original system or not. The parameters that need to be observed are the terms of the separation between aircraft whether it has been fulfilled. The model also has not been able to accommodate the characteristics of the relationship between the pilot, controller and ATC system so that it is still necessary to improve the simulation model at the next research stage.

Additional data is also still needed regarding the guiding characteristics carried out by the controller, especially when facing critical situations. The interaction between humans and the ATC system also requires further exploration through in-depth interviews with the controller.

## 4. Conclusion

This article discusses the progress of research on developing a hybrid simulation model of an ATM system in the Terminal Control Area. Research has developed an agent-based ATM system simulation model at the Soekarno-Hatta Airport Area Terminal which is equipped with visualization of aircraft movements. Simulation model provided with visualization is an effective method for representing and analyzing complex systems such as ATM systems. However, validation of the simulation model is still needed to prove the simulation model developed can represent the ATM system accurately. The simulation model cannot be used to adequately reflect the characteristics of the ATM system in the Terminal Area because it does not include interactions between the pilot, controller and ATC system. Improvement of the simulation model is still needed to represent the socio-technic system better.

**Acknowledgments**
This research was supported by the Ministry of Research, Technology and Higher Education of the Republic of Indonesia within the Penelitian Terapan Unggulan Perguruan Tinggi (PTUPT) Research Program.

**References**
[1] ICAO 2016 Asia / Pacific Seamless ATM Plan
[2] ICAO 2017 *Asia/Pacific Framework for Collaborative Air Traffic Flow Management* (Bangkok)
[3] Boeing 2017 *Current market outlook 2017-2036* (Seattle, WA: Boeing Commercial Airplanes Market Analysis)
[4] The Jakarta Post 2017 Indonesia's air traffic controllers warn of collision risks as flight volume rises
[5] Bouarfa S 2015 *Agent-Based Modelling and Simulation of Safety and Resilience in Air Transportation* (Delft University of Technology)
[6] Wei Z and Li J 2015 Design of Air Traffic Control Operation System Using Multi-Agent Technology and Simulation *Open Fuels Energy Sci. J.* 368–74
[7] Fakhimi M and Mustafee N 2012 Application Of Operations Research Within The Uk Healthcare Context *Proceedings of the Operational Research Society Simulation Workshop 2012 (SW12)* vol 2012, ed and D-J van der Z B.Tjahjono, C. Heavey, S. Onggo
[8] Glover W and Lygeros J 2004 A Stochastic Hybrid Model for Air Traffic Control Simulation *International Workshop on Hybrid Systems: Computation and Control* pp 372–3
[9] Pasaoglu C, Baspinar B, Ure N K, and Inalhan G 2016 Hybrid systems modeling and automated air traffic control for three-dimensional separation assurance *Proc. Inst. Mech. Eng. Part G J. Aerosp. Eng.* **230** 1788–1809
[10] Janic M 2000 *Air Transport System Analysis and Modelling* (Florida, USA: CRC Press)
[11] Neufville R De and Odoni A 2003 *Airport Systems: Planning Design, and Management* (McGraw Hill)
[12] Endsley M R 1995 Toward a Theory of Situation Awareness in Dynamic Systems *Hum. Factors J. Hum. Factors Ergon. Soc.* **37** 32–64
[13] Bouarfa S, Blom H H A, Curran R and Everdij M H 2013 Agent-based modeling and simulation of emergent behavior in air transportation *2013 Aviat. Technol. Integr. Oper. Conf.* **1** 1–16

IOP Conf. Series: Materials Science and Engineering **645** (2019) 012004 doi:10.1088/1757-899X/645/1/012004

# Experiment of Thermal Insulation Performance of Acrylic and Expanded Polystyrene for Chamber Wall

**Q Laliyah[1,*], P Bakti[1], and W S Ramadhani[1]**

Research Center for Quality System and Testing Technology Indonesian Institute of Science, Puspiptek Region Building 417, Setu, Tangerang Selatan, Banten, Indonesia - 15314

[*]E-mail: quds001@lipi.go.id

**Abstract**. Chamber with the controlled temperature is required in the energy efficiency test of a household product. Chamber wall plays an important role to keep the temperature inside the chamber so as not to be affected by the temperature from outside. Experiments were carried out to compare the thermal performance of two different material for chamber wall : 1) Acrylic and 2) expanded polystyrene. The boxes with dimension 40 x 40 x 40 cm were affected with fixed thermal loads in a climatic chamber. Meanwhile, temperature inside the box was monitored with data logger. The thermal loads for testing were 40°C, and 60°C. Based on the measurement, temperature in the acrylic box reach the load temperature from the outside and starts steady at 300 seconds, while the expanded polystyrene at 550 seconds. The result show that the insulating performance of expanded polystyrene is better than acrylic. The result of this study can be used not only for refrigerator manufacture, it can also use by another industry related to thermal dynamic. The industries include aeronautic industry related to aircraft wall coatings for external temperature protection, building industry in terms of building wall coatings to maintain temperature stability, and other related industries.

## 1. Introduction

In Indonesia, according to electricity statistics 2016 the use of electrical energy for the household sector ranks first at 43.35% followed by the industrial sector 31.5%, the business sector and the public respectively 6.55% [1]. The dominant use of electricity in the household sector is partly due to the increasing number of households from year to year. In addition, the increase in middle-class economy households has led to increasing use of electricity-based household appliances including air conditioners, ballast lamps, water pumps, refrigerators, fans, televisions, rice cookers, irons and washing machines. Responding to these conditions, it is deemed necessary to conserve energy. One of the efforts to support energy conservation is by implementing mandatory standards for energy labeling for household electrical appliances.

Refrigerator is one of the household electrical appliances which is the biggest contributor to the use of electrical energy, especially for the price class R1-450VA and R1-900VA [2]. In 2013 the Indonesian government through the BRESL program had targeted the labeling of energy for refrigeration products [3]. To support the labeling energy program, it is necessary to prepare the test laboratories in developing standard testing methods to provide a standard guide for uniform testing methods and can be repeated to measure the characteristics of an equipment [4]. Especially for refrigerators, the international standard referred to the energy consumption testing process is IEC 62552: 2015, Household Refrigerating appliances - Characteristics and Test methods. Based on these standards [5], testing the refrigerator's energy consumption must be carried out in a room with an ambient temperature of 32°C + 0.5°C (for tropical regions).

In order to fulfill the energy efficiency testing procedure of refrigerator, it is necessary to provide chamber with controlled temperatur. Chamber wall plays an important role to keep the temperature inside the chamber so as not to be affected by the temperature from outside. Miskinis et al [6] presented in their paper how (expanded polystyrene and mineral wool) thermal insulation and coating (plaster) layers of ETICS (External Thermal Insulation Composite System) influence on sound and thermal insulation of a basic element (wall). The experimental results showed that added mineral wool thermal insulation layer of ETICS increases sound insulation up to 4dB, but in using EPS (Expanded Polystyrene) has no significant influence. The addition of a coating (plaster) layer over the thermal insulation material layer allows an increase in sound insulation in both cases from 3 dB to 7 dB, but does not change the thermal insulation. The addition layer of EPS thermal insulation increases thermal insulation of the basic wall up to 4.88 $m^2K/W$ and up to 5.09 $m^2K/W$ with mineral wool. The influence on sound insulation depends on the type of thermal insulation material used, the presence or absence of a coating (plaster layer) and its thickness. While the influence on thermal insulation only depends on the presence of the thermal insulation layer. Khoukhi [7] investigated the effect of the operating temperature on the thermal conductivity of polystyrene insulation material using the developed experimental apparatus based on the guarded hot plate principle. The result showed that the k-value is affected by the change in operating temperature. The result also showed that the lower material density is, the higher is the thermal conductivity.

In this study researchers tried to find out the thermal performance for acrylic and expanded polystyrene, the selection of both materials is based on the ease of getting it, cheap, and easy to install. Measurements in this study are limited to the material's response of heat from outside. From these measurements it is expected to be able to know which insulator material is better to be applied in the manufacture of a refrigerator energy efficiency test chamber. The result of this study can be used not only for refrigerator manufacture, it can also used for another industry related to thermal dynamic. The industries include aeronautic industry related to aircraft wall coatings for external temperature protection, building industry in terms of building wall coatings to maintain temperature stability, and other related industries.

## 2. Material and Method

### *2.1. Material*

Acrylic with a thickness of 2 mm and expanded polystyrene with a thickness of 5 cm was selected in this research. Acrylic and polystyrene glue needed to make the boxes. The sample test used is boxes from acrylic and expanded polystyrene with dimension 40 x 40 x 40 cm.

To measure temperature of boxes, a type K of thermocouple with wire diameter of 0.3 mm was used. Meanwhile, data logger used to monitoring the temperature. Measurement were carried out in a climatic chamber to control the ambient temperature.

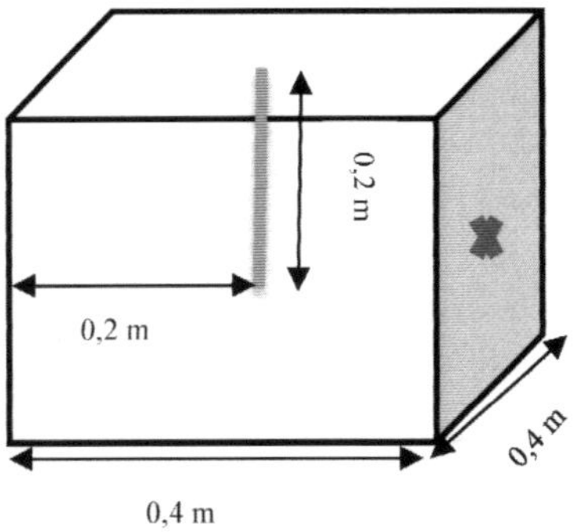


**Figure 1**. Sample box with dimension 40 x 40 x40 cm

### *2.2. Method*
Experiments were carried out to compare the thermal performance of two different material for chamber wall: 1) Acrylic and 2) expanded polystyrene. To identify thermal performance the boxes with dimension 40 x 40 x 40 cm were affected with fixed thermal loads in a climatic chamber. Meanwhile, temperature inside the box was monitored with data logger. The thermal loads for testing were 40°C, and 60°C. Sample boxes are inserted into the chamber when temperature of chamber already constan as same as a temperature load. A thermocouple type K for measure the temperature installed on the center of sample boxes and on center of outer and inner wall chamber as shown in Figure 1. Measurements were carried out for 3600 seconds. Temperature changes can be seen from the graph of measurement results. Based on the resulting graph can be analysed of thermal performance of each insulator material.

## 3. Results

### *3.1. Ambient Temperature*
The ambient temperature during the measurement, which lasted for 3600 second, is shown in Figure 2.

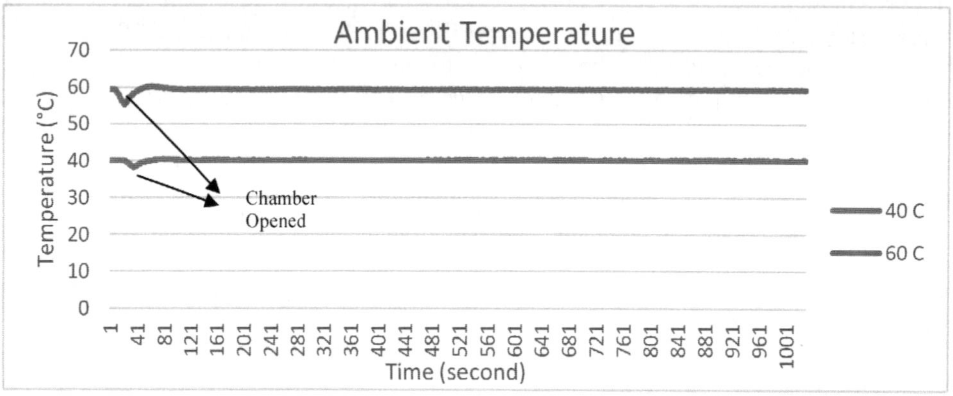


**Figure 2**. Graph of ambient temperatur inside chamber during the measurement

Figure 2 shown that the ambient temperature have a small fluctuations during constan temperature period on the regulation system of the climate chamber. The regulation was on during all the measurement. This temperature profile i chosen to represent the temperature required by the refrigerator efficiency energy test standard [5,8].

### *3.2. Acrylic Temperature Change*
Thermocouple are installed on an acrylic box on three different places. The temperature on acrylic box was monitored with data logger.

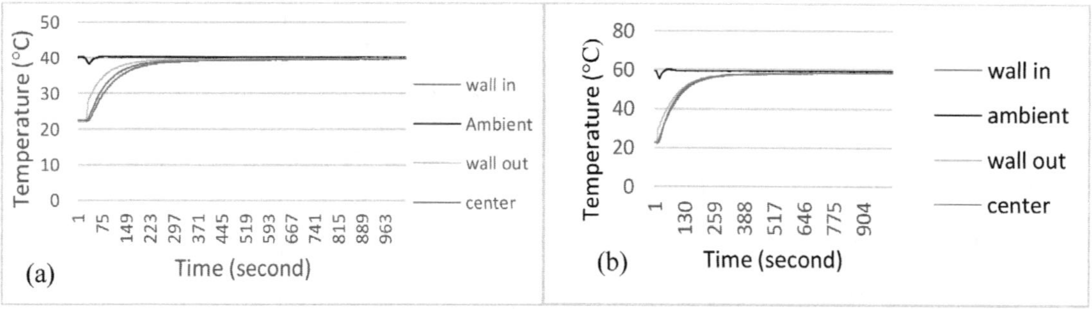


**Figure 3**. Graph of Temperature in three different places on acrylic (a) at 40°C, (b) at 60°C

Temperature inside the acrylic box both on a wall and in the center of the box are shown in Figure 3. The graph shows that both on 40°C and 60°C, external heat can propagate quickly into the box through an acrylic wall. When the thermal load is 60°C, the heat propagation takes a short time to reach steady state at 60°C. Judging from the results, acrylic cannot hold thermal load from the outside properly.

### *3.3. Expanded Polystyrene Temperature Change*

Same as in acrylic box, thermocouples are installed on the expanded polystyrene on three different places. The monitored temperature as shown in Figure 4.

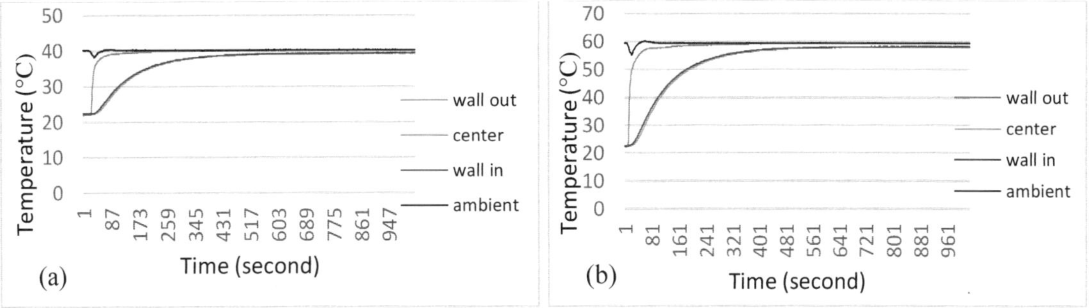


**Figure 4**. Temperature graph in three different places on expanded polystyrene (a) at 40°C (b) at 60°C

Similar to the results on acrylic, figure 4 show that heat propagation in expanded Polystyrene requires less than 3600 seconds to reach steady state. Judging from the Figure 4, there is a difference in the heat transfer time on the outer wall and inside of the box. Outer wall of expanded polystyrene has a rapid rise in temperature before steady state. Meanwhile, temperature inside the box rises slowly before reaching steady state.

Comparison between Figure 3 and Figure 4 reveals that the difference between two insulator materials is not significant because the steady state conditions are achieved in less than 3600 seconds and the steady temperature is the same as the ambient temperature. However, from the graph in Figure 5 show that the heat propagation on the expanded polystyrene wall takes longer to get the steady state at the same temperature as the ambient temperature.

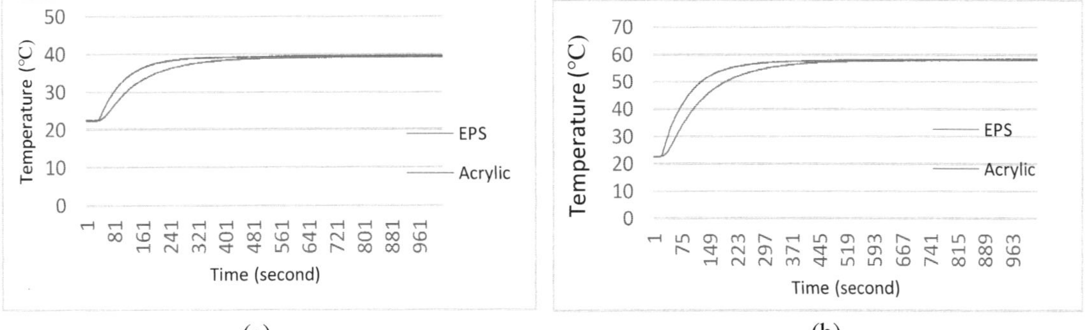


(a) (b)

**Figure 5**. Graph of Temperature on the center of boxes (a) at 40°C, (b) at 60°C

Wall of thickness and average thermal conductivity are effect to heat transfer. The thicker wall will reduce heat transfer, while the greater the thermal conductivity, the greater the heat transfer will occur [8]. Based on the graph of measurement results, expanded polystyrene can hold the heat better than acrylic, this can be caused by three factors. Firstly, due to the expanded polystyrene sample used has a thicker wall thickness. Secondly, expanded polystyrene has a lower thermal conductivity than acrylic. Thermal conductivity of expanded polystyrene is 0,03 W/m.K and thermal conductivity of acrylic is 0,2 W/m.K [9]. Thirdly, expanded polystyrene contains up to 98% of its volume of tiny closed bubbles of air which inhibit the passage of heat or cold [10].

There is a study conducted by Lakatos (2014) related to this study. The material used is expanded polystyrene with different qualities, namely EPS 30, EPS 100, EPS 150, and EPS 200. The result of that study is that the longest delay can be reached with the use of EPS 200. It has relatively big material constants and small thermal conductivity compared to other insulators. So, in order to reach a long retardation time, small thermal diffusion coefficient is needed. It can be reached by relatively small thermal conductivity and with big material constants [11].

## 4. Conclusion

Thermal performance of materials insulation for chamber wall has been investigated in this study. Based on the measurement, insulating performance of expanded polystyrene slightly better than acrylic. Wall of thickness, average thermal conductivity of materials and its closed cell structure which filled with air are effect to the thermal performance of insulation material.

## References

[1] Electrical Power Statistics 2016. Directorate General of Electricity of the Ministry of Energy and Mineral Resources. Edition No. 30 Year Budget 2017.

[2] Energy Efficiency and Elasticity Planning 2012. Balai Besar Teknologi Energi, Badan Pengkajian dan Penerapan Teknologi

[3] http://www.id.undp.org/content/indonesia/en/home/operations/projects/environment_and_energy/barriers-removal-to-the-cost-effective-development-of--energy-ef.html accessed on September 29nd 2017 at 4.16pm

[4] Mahlia T M I, Masjuki H H and Choudhury I A 2002 Theory of energy efficiency standards and labels *Energy Convers. Manag.* **43** 743–761

[5] International Electrotechnical Comission (IEC) 2015. IEC 62552-1 Household Refrigerating appliances - Characteristics and Test methods

[6] Miskinis K *et al* 2018 Influence of EPS, mineral wool and plaster layers on sound and thermal insulation of a wall: a case study *Applied Acoustic* **137** 62-68

[7] Khauki M 2016 Change of the Thermal Conductivity of Polystyrene Insulation in term of Temperature at the Mid Thickness of the Insulation Material: Impact on the Cooling Load. Word Academy of Science, Engineering and Technology *International Journal of Mechanical and Mechatronics Engineering* **10** 5

[8] Yunus A Cengel. Heat Transfer A Practical Approach. *Textbook.* Second Edition

[9] The Engineering Toolbox. Thermal conductivity of Common Material and Gases. https://www.engineeringtoolbox.com/thermal-conductivity-d_429.html accesed on September 15nd 2018 at 10.20pm

[10] Poletto M Junior L O and Zattera A J 2014 Expanded Polysterene: Thermo-Mechanical Recycling, Characterization and Application in Polystyrene: Synthesis, Characteristics and Applications, New York, Nova Publishers 53-74

[11] Lakatos A 2014 Comparison of the Thermal Properties of Different Insulating Materials *Advanced Materials Research* **899** 381-386

# Preliminary Design of GL-1 Sailplane Cockpit Arrangement and Control Mechanism Based on EASA CS-22 Requirements

**T Mulyanto[1,*], M F Zulkarnain[1]**
[1] Faculty of Mechanical and Aerospace Engineering, Institut Teknologi Bandung, Jl. Ganesha No. 10, Bandung, Indonesia.
*Email: taufiq.mulyanto@ae.itb.ac.id

**Abstract**. Sailplaning or gliding is one branch of aero sports activities in Indonesia under Federation of Aero Sport Indonesia (FASI). Until now, the most common sailplane used in Indonesia is still Schweitzer SGS 1-26, which is a single seated metal-fabric sailplane designed in 1955 and produced until 1979. The development of a new national glider to replace the Schweitzer SGS 1-26 had been conducted since 2013 and resulted in a design named GL-1. The EASA CS-22 had been chosen as certification basis for GL-1 as there is no equivalent yet in Indonesian CASR. A preliminary design of internal cockpit arrangement and control mechanism has been conducted. This study shows that the control mechanism using a combination of cable and push-pull rod arrangement satisfy the required control surface movement and limited sailplane interior space.

**Keyword**: Sailplane, Preliminary Design, Certification, Cockpit, Control Mechanism.

## 1. Introduction

Preliminary study on mid-performance glider were conducted since 2013 with some configuration study [1] and performance estimation [2] as a study final report. However, GL-1 was first designed by Pratama [3]. Estimation of GL-1 aerodynamic characteristics and performances had been explored by using analytical method [2], drag polar method based on flight manual data [3][4], CFD simulation [5], and X-Plane simulation [6]. Amalinadi [7] had also studied distribution of aerodynamics forces over GL-1 wings using CFD. A half scale prototype called BL-1 based on the design by Pratama was built in 2016. It was displayed on Indonesian National Glider Workshop that was held during biannual National Sport Event. BL-1 had conducted a wing static test in 2016 and flight test at April 2017 and May 2017 [8]. Structural design and analysis had been conducted by Darsono [9]. The full scale prototype is in production and is targeted to have its first flight in 2019.

To continue the development and achieve 2019 flight test, design of the internal configuration, especially in cockpit and control mechanism, will be needed. This paper will explain cockpit and control mechanism design process and results. European EASA CS-22 is used as basis requirements as there is no equivalent regulation in Indonesia. The study will be focused on the configuration and arrangement of seating, cockpit control, instrument panel, and control mechanism. For control mechanism, only elevator, aileron, and rudder will be discussed. Other related to cockpit design such as stick force, structural integrity, emergency condition, etc will be discussed separately.

## 2. GL-1 Design and Specifications

GL-1 is a single seater mid performance sailplane. Its low rate of sink with moderate aspect ratio is designed for Indonesian small and weak thermal condition. GL-1 is also designed with low stall speed so the aircraft could be towed by microlight or ultralight, which is more common in Indonesia

IOP Conf. Series: Materials Science and Engineering **645** (2019) 012005 doi:10.1088/1757-899X/645/1/012005

compared to tow aircraft and winch launcher. Aircraft specification and dimension can be seen in Table 1 and Fig. 1.

**Table 1**. Summary of GL-1 technical data

| | |
|---|---|
| Wing Span | 14.3 m |
| Wing Area | 12 $m^2$ |
| Wing Aspect Ratio | 17 |
| Fuselage Length | 6.8 m |
| Empty Weight | 190 kg |
| MTOW | 300 kg |
| Best L/D | 30 |
| Minimum R/D | 0.76 m/s |
| Stall Speed | 14.8 m/s |
| Elevator Deflection | $25^0$ up and $30^0$ down |
| Aileron Deflection | $30^0$ up and $10^0$ down |
| Rudder | $25^0$ right and left |
| Flap | $25^0$ down |

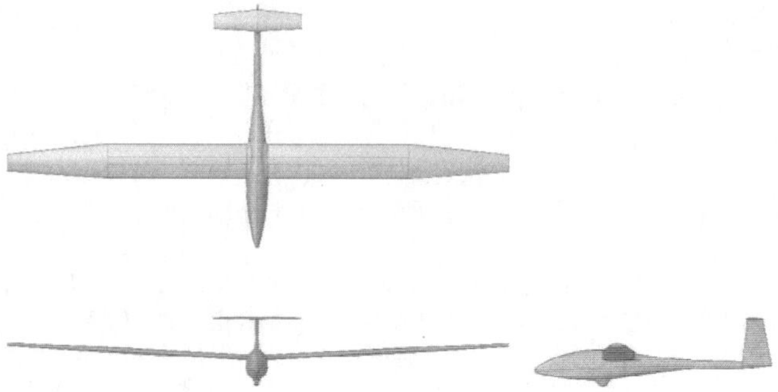

**Figure 1.** GL-1 Three View Drawing

## 3. Certification Study

EASA CS-22 is certification specific for sailplane and motorized sailplane [10]. The structure of CS-22 is actually very similar to CASR 23 [11], with special consideration for sailplane design such as towing, water ballast, etc.

For this design, specific regulation for mechanism and cockpit will need to be collected. There are total of 32 regulation for mechanism and cockpit. Most of arrangement, design features, and operation are discussed primarily in Subpart D of CS-22. Handling quality and performance are discussed in Subpart B and D. While equipment lists are regulated in Subpart F. However, for the scope of this study, only requirements that related to configuration and arrangements are considered for preliminary design.

### *3.1. Cockpit Requirements*

CS 22.771 and .777 explained that the cockpit design and arrangements must be designed to able the pilot to control the aircraft accordingly. The pilot must be able to move the control freely without any interference from any objects in the cockpit. Arrangements and operation of the cockpit control must be designed to avoid confusion and inadvertent operation. CS 22.779 and CS 22.780 regulate the arrangement and operation of the main cockpit controls. However, the requirements for instrument panel is not as clear as in CASR-23, so a study on typical instrument arrangements on similar sailplane must be conducted.

### *3.2. Control Mechanism Requirements*

As stated in CS 22.671 and CS 22.685, the operation and arrangement of control mechanism must be easy, smooth, and positive, avoiding confusion and inadvertent operation. As explained in 3.1, the operation of control mechanism is regulated on CS 22.779 and CS 22.780. The design must also conform to proper maximum control load and convenient stick force direction. Furthermore, the mechanism must be able to operate without interference from other mechanism or structural object. This was checked using analytic and kinematics simulation.

In this study, the aircraft is assumed as rigid body. However, it is not the case in reality. Usually, high performance gliders have a hogh deformation, and the study on the effect of those high deformation to control mechanism must be conducted separately.

## 4. Results

*4.1 Cockpit Design*

The cockpit of GL-1 is designed for Indonesian pilot with height between 1.6 meters to 1.85 meters [12]. The cockpit design geometry variable were pilot location, seat inclination, and knees location. For light sport cockpit [13], the pilot is usually located in the aft area of the cockpit due to wider cross section area. The inclination of backrest will usually be small and knees and leg require to be wide and provide comfortable seating position. However, due of the very narrow design of the GL-1 nose and cockpit area, a careful compromise will be needed. Narrowing front fuselage will make the pilot moved forward and have a considerable amount of back inclination. The leg area will also be cramped especially in the area of the rudder pedal.

After several iteration, the pilot is positioned 1.38 meters from nose with 40 degrees seat inclination. The location is designed so it will not be too cramped for the tallest pilot and the control is still reachable for shortest pilot. Figure 2 shows the pilot seating position.

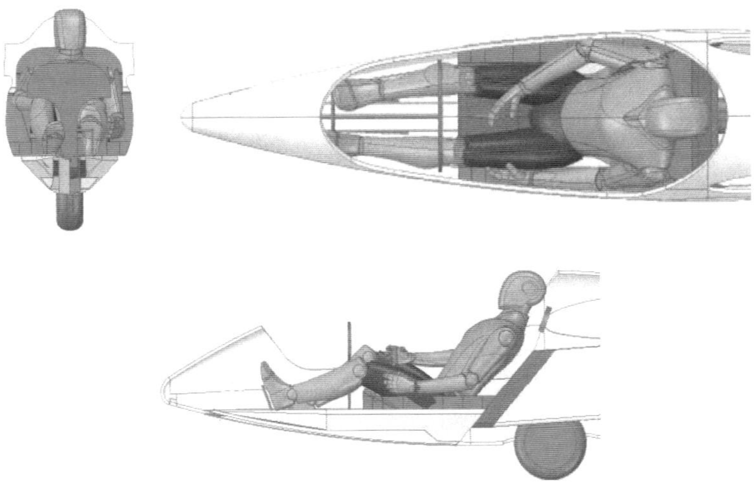

**Figure 2.** Pilot Location and Seating Configuration

Beside CS-22 requirements, existing sailplanes type in Indonesia is also studied. A familiar cockpit arrangement would ease pilot transition to the new sailplane. Some arrangement highlights were made. Control stick is located in the middle between pilot feet. Airbrake and flap handles are located on the left-hand side of the cockpit so the right hand will always be on stick during flight. Tow release handle is located to the left side below the stick. Adjustable rudder pedal is installed to accommodate wide range of pilot height.

There is no clear guidance in CS-22 on arrangement of instruments panel. According to the CS 22.1303, the instruments needed for normal flight are Airspeed Indicator and Altimeter. However, existing glider usually adds more instruments such as Variometer and GPS. Thus, a study of more than 40 gliders from year 1970 to modern glider with varied role and performance was conducted to analyze the trend of instruments arrangements in glider. A list of instruments type will be listed and counted. For each type of instrument and each glider a tabulation of size and location were made. The GL-1 instrument type and arrangement were decided based on the statistic from the tabulation. A study on eye-instruments distance is also conducted from several existing gliders.

Other than airspeed and altimeter, variometer and magnetic compass are always installed in most glider. Another different type variometer sometimes added in the panel, so there will be two variometer. A radio/atc transceiver and a display/gps/computer is usually installed on more than 70% selected glider, especially on glider made after 1990. Attitude indicator instruments is rarely used, counted only 40%, however this is compensated as every glider usually have sideslip indicator in the canopy. Thus, all of the instruments listed which are frequently used will be installed in GL-1. A turn-bank indicator will also be installed as it is frequently used on other light aircraft.

Instrument panel is usually divided into several section horizontally and vertically: above instruments, upper left, upper center, upper right, far left, mid left, mid center, mid right, far right, bottom left, bottom center, bottom right, and below instrument panel. The arrangement of airspeed, altimeter, an attitude indicator is usually very similar to CASR-23: airspeed left, attitude center, and altitude right. The radio/atc transceiver is always located on the bottom center part of the panel. Since there are usually two variometers, their location is usually separated and placed on the remaining empty space. Magnetic compass is usually located above the panel, on the right of panel, or outside the panel.

Instruments are usually come with the circular size of 3” and 2”. However, radio and LCD display have rectangular form. In modern glider, the use of large display for performance computer or GPS is on the rise. Due to the size of the display, the other instruments will need to be shrink in size. Instrument type which usually installed with 3” size in older gliders, are all installed in 2” size for modern glider with computer or GPS display. Even though the instruments are smaller, the eye-instruments distance is larger. The trend shows that modern gliders have distance of 600-750 mm compared to 550-750 mm in 1970s. Thus, the farthest distance will be used, but with a balance of 3” and 2” instruments for clear pilot view and minimize panel space as no large display will be installed. The large display for computer or gps can be replaced by adding 3” computer/gps instruments or by adding attachments for external gps using smartphone.

The instrument panel will be located 930 mm from nose to accommodate maximum eye distance of 750 mm. The panel will be designed so that enough instruments could be installed. There are also leg space so the movements of pilot’s leg will not be inhibited during normal operations and emergency.

As depicted in figure 3, there are 8 slots in the instrument panel with four 3” slots and four 2” slots. There are three slots in the upper part, four slots in the middle, and one slot in the bottom. A 3” Airspeed Indicator, a 2” Attitude Indicator, and 3” Altimeter will be located in the upper left, upper center, and middle right respectively. A 3” Vertical Speed Indicator or Variometer will be located in the upper right. There is three possible locations for magnetic compass, above the instrument panel, on the far right of the instrument panel, or on the cockpit wall. A 2” slot in the far left is provided for a Handheld/Mobile GPS adaptor. An extra Variometer, a Performance Computer, or a GPS could be installed in the 3” slot in middle left.

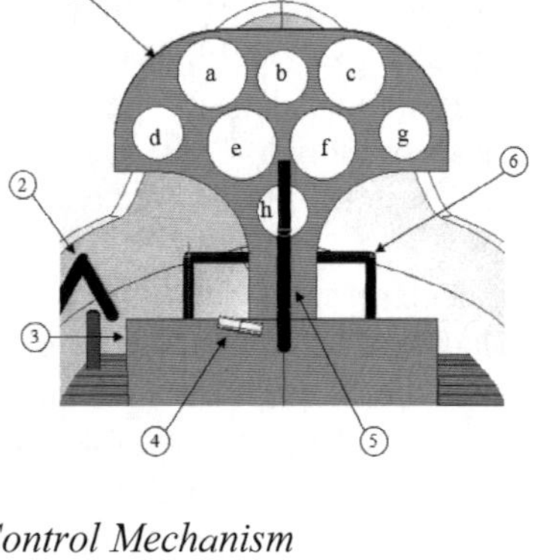


**Figure 3.** Cockpit Arrangements of GL-1: 1) Instrument Panel, 2) Flap, 3) Airbrake, 4)Tow Release, 5) Control Stick, 6) Rudder
Instruments Arrangements of Gl-1: a) Air Speed Indicator, b) Turn/Bank Indicator, c) Vertical Speed Indicator, d) GPS Adaptor, e) VSI/Computer/GPS (optional), f) Altimeter, g) Magnetic Compass, h) Transceiver (optional)

*4.2 Control Mechanism*

As there is no reference for configuration of glider control mechanism, a study of 11 maintenance manual, one glider drawings, and light sport aircraft construction handbook was conducted [13].The construction of light or sport plane control mechanism usually use steel cable. This is reflected on 1950s glider Schweizer 1-26 [14], which is the main glider operated in Indonesia [15]. However, this is not the case on the more modern sailplane such as PW-5 [16], Arcus [17], or LAK-17 [18]. The more modern and high-performance sailplane used combination of push-pull, torque tube, and crank as the main components for control mechanism. Steel cable use is limited to rudder and tow release mechanism. This might be the case as mechanism using push pull rod will have positive and smoother control [13]. This in line with the requirements explained in 3.2. Thus, GL-1 will use this design.

A study on existing glider control mechanism configuration and arrangements was conducted and some features are used as the main reference for design. Other consideration is to minimize the complexity of the mechanism aiming to ease manufacture process and reduce cost. Configuration which fulfill requirements explained in 3.2 and maximizing the ojectives was chosen as the final configuration. However, one cannot simply transfer an existing control mechanism to fit in other glider since there are difference in geometry and other constraints. The chosen basic control mechanism configuration will need to be designed so that the control is within reach of the pilot, while the full mechanism has no interference with pilot, structures, or other control mechanism. Control mechanism operation requirements in CS 22.779 and 22.780 must also be fulfilled. The control must have a tendency to return to its neutral position, thus requiring the stick force direction must be opposite of the control movements. In some cases, a combination of two basic control mechanism configuration and a considerable design adjustment were need.

Control column in the middle was chosen as it is more familiar with Indonesian Pilot. Movement transfer for elevator and aileron passed through the sides of the cockpit. Even though control passage through the middle of the fuselage could provide simpler design, it wasn't possible since the space are very limited due of the existence of tow mechanism and landing gear. Figure 4 shows GL-1 control mechanism configuration and arrangements.

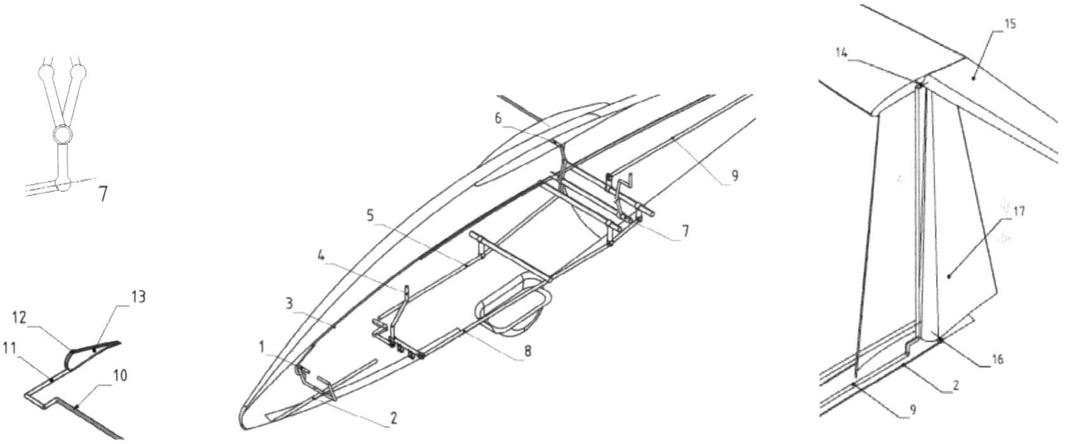


**Figure 4.** Control Column. Aileron Differential (top left), Wing Mechanism (left), Fuselage Mechanism (middle), and Empennage Mechanism (right).
Legend: 1) Rudder Pedal, 2) Adjustable Rudder Mechanism, 3) Rudder Cable, 4) Control Stick, 5) Aileron Fuselage, 6) Aileron Fuselage to Wing, 7) Aileron Differential, 8) Elevator Fuselage, 9) Elevator to Empennage, 10) Aileron Wing, 11) Aileron Horn, 12) Aileron Hinge, 13) Aileron, 14) Elevator Horn, 15) Elevator 16) Rudder Horn 17) Rudder

The elevator uses push pull rod mechanism. The movement transfer is designed so when the pilot pulls the stick, the elevator will be deflected upwards. A modified LAK-17 elevator mechanism [17] was chosen as basic configuration for it simple mechanism. The main modification is to replace middle crank so the rod end bearings terminal that was used will be replace with simpler and cheaper fork end terminal.

The aileron is a push pull rod mechanism. It is designed so when the stick is rotated to the right, the right aileron will be deflected upwards. There are three gliders with distinct aileron mechanism configuration: Schempp Hirth series [16], DG Series [18] and LS series [19]. However, since they all have a combined flap-aileron, a modified Schemmp Hirth mechanism was chosen as basic configuration because it's easier to convert into aileron-only mechanism. The design also incorporated a feature to enable different aileron deflection between right and left aileron.

The rudder has a steel cable mechanism. It was designed so that when right rudder is deflected forward, the rudder will be deflected to the right. The rudder pedal is actuated by leaning/rotating the pedal forward instead of pushing the rudder pedal forward. This is needed to implement an adjustable rudder mechanism.

## 5. Conclusion

The preliminary design of GL-1 cockpit and control mechanism had been conducted. From 32 articles in CS-22 related to control and cockpit design, only nine related to arrangements and configuration is considered. The study of more than 40 glider, 11 maintenance manual, two drawings, and a light sport construction handbook was conducted to design the configurations and arrangements that could fulfill all the requirements. Instrument panel has 8 slots, two of them could be used to accommodate Handheld GPS, Performance computer with LCD display. Control mechanism used a combined cable and push-pull rod arrangement which can satisfy the required control surface movement and limited sailplane interior space.

## Acknowledgement

This research was partially supported by P3MI ITB research scheme.

## References

[1] Jan Sulc 2013 *Conceptual Design of East Asia Sailplane* Institut Teknologi Bandung Design Final Report
[2] Vega Handojo 2013 *Sailplane Performance Estimation* Institut Teknologi Bandung Student Project Report
[3] Hendi Pratama 2015 *Conceptual Design of Indonesia National Glider GL-01* Institut Teknologi Bandung Master Thesis
[4] Azizah, N. A. D., Nurhakim, M. L. I., Arifianto, O., Mulyanto, T. 2016 Glider Uncentered Drag Polar Estimation Based on Flight Manual Data. *Advances in Aerospace Science and Technology in Indonesian Vol.1 2016*
[5] Amalinadhi, C., Julistina, R. 2017 *GL-1 CFD Simulation Result* GL-1 Development Archives
[6] Zulkarnain, M.F. 2017 *GL-1 X-Plane Simulation Result* GL-1 Developments Archives
[7] Amalinadhi, C., Julistina, R., Moelyadi, M. A., Mulyanto, T. 2016 Comparative Study Between Schrenk and CFD Analysis for Predicting Lift Distribution Along Wing Span of Glider Aircraft. *Advances in Aerospace Science and Technology in Indonesian Vol.1*
[8] Zulkarnain, M.F, Rahman, M.F., Nurhakim, M.L.I, Arifianto, O., Mulyanto, T. 2017 Flight Test of GL-1 Glider Half Scale Prototype *5th Int. Seminar of Aerospace Science and Technology*
[9] Darsono, R 2017 *Desain dan Analisis Struktur Sandwich Fuselage Pesawat Glider GL-1 Berbasis Metode Elemen Hingga dan Analitik* Institut Teknologi Bandung Bachelor Thesis
[10] EASA 2009 *Certification Specifications for Sailplanes and Powered Sailplanes CS-22*
[11] Ministry of Transportation 2014 *Civil Aviation Safety Regulation Part 23*
[12] Kroemer, K.H.E., Kroemer, H.B. and Kroemer-Elbert, K.E. Ergonomics 2001 *How to Design for Ease and Efficiency*
[13] Tony Bingelis 1992 *Sportplane Construction Techniques A Builder's Handbook*
[14] *Schweizer 1-26B Assembly Drawing*
[15] Pordirga Terbang Layang 2007 *Buku Pedoman Terbang Layang Indonesia*
[16] 2001 *Sailplane Maintenance Manual B1-PW-5*

[17] Schempp Hirth Flugzeugbau GmbH 2012 *Maintenance Manual for Powered Sailplane Arcus M*
[18] Sportine Aviacija 2006 *Maintenance Manual for the Self-Sustaining Powered Sailplane LAK-17AT*
[19] DG Flugzeugbau GmbH 2009 *Maintenance Manual DG-500*
[20] Rolladen-Schneider Flugzeugbau GmbH 1999 *Maintenance Manual for the LS-8a Sailplane*

# Aerodynamics performance of glider GL-1 based on computational fluid dynamics in optimum flight conditions

**E Amalia[1], M A Moelyadi [1], R Julistina[1]**

[1]Faculty of Mechanical and Aerospace Engineering, Institute of Technology Bandung, Jl. Ganesha 10 Bandung 40132, Indonesia

Corresponding author: E Amalia (ema@ftmd.itb.ac.id)

**Abstract.** Glider GL-1 is designed as a national glider of Indonesia for aero sport gliding activity and fall into category of 15 meters class by FAI (world's air sport federation). In preliminary design phase, design requirements are derived from thermal updraft condition of Indonesia which lead to maximum rate of descent of 3 m/s and maximum turning radius of 150 m. GL-1 has wing area of 12 $m^2$ and wing aspect ratio of 17. Wing span of GL-1 is 14.283 m with length of fuselage is 6.795 m. Performance study in preliminary design gives two optimum flight conditions for GL-1 to glide, which is maximum range condition and maximum endurance condition. In maximum range condition, with maximum aerodynamics efficiency between 24 to 30, GL-1 has a velocity of 20 – 25 m/s with density of 1.225 kg/m3. In maximum endurance condition, GL-1 has a velocity of 17 – 20 m/s with the same density and minimum rate of descent between 0.66 – 1.1 m/s. Reynolds number is between 1 – 1.5 million. This paper present CFD simulation result for both optimum flight conditions predicted in preliminary design that carried out at velocity of 25 m/s for maximum range condition and 17.5 m/s for maximum endurance condition. By using full-glider meshing with total element of about 11 million, a maximum aerodynamics efficiency of 21 is obtained for maximum range condition with lift coefficient of 0.91. While minimum rate of descent for maximum endurance condition of 0.87 m/s is obtained.

## 1. Introduction

Glider GL-1 was firstly designed with the name of GL-01 or "Gajah Layang 01", and reference [1] documented the conceptual design phase of the glider and currently the design process has entered detail design phase. This national glider is designed in cooperation with FASI (Indonesia Aero Sport Federation) to answer problem they face. The problem was FASI does not have a glider that suitable for thermal updraft condition in Indonesia which is narrow and weak. Configuration of glider GL-1 is a single seated high-wing airplane with T-tail with span of 14.283 m , MAC of 0.863 m, wing area of 12 $m^2$, inner wing taper ratio of 1, outer wing taper ratio of 0.5, and dihedral of 3 degree.

Derived from thermal updraft condition in Indonesia that having diameter between 200 to 300 m, maximum height about 5000 – 6000 ft, and updraft strength of 0.5 – 3 m/s, we get performance requirement for glider GL-1 which are maximum rate of descent of 3 m/s and maximum turning radius of 150 m. This leads to a minimum aerodynamics efficiency of 8.333 for a condition of non-optimum flight with calculation based on reference [2]. Moreover, some preliminary calculation of performance of glider GL-1 has been done for optimum flight conditions derived from thermal updraft characteristics of Indonesia as in reference [1],[3], and [4].

Two optimum flight conditions arc maximum range with maximum aerodynamics efficiency ($C_L/C_D$) and maximum endurance with minimum rate of descent. For maximum range flight condition, by using national aerosport competition rule, we take the rule for short flight which is the release altitude is 1000-2000 ft with a condition of ready to land when in an altitude of 500 ft. Then, we take a release height for GL-1 to fly is within 1500 ft and using gliding symmetric flight method in reference [2]. While for maximum endurance flight condition, we take the rule for endurance with a release height for GL-1 to fly is within 1500 ft with using thermal updraft as much as possible. So, for maximum endurance flight condition, we use cross country fligth method in reference [2]. Here, we use an average value of thermal updraft vertical speed of 1.75 m/s. Table 1 shows performance prediction of GL-1 from preliminary design phase for the two optimum flight conditions compared with results of current study. We will discuss it later in section 4.

Recently, glider GL-1 has entered detail design phase, therefore a detail aerodynamics performance analysis by using CFD (Computational Fluid Dynamics) is needed. Evaluation by using CFD presented in this paper will cover the two optimum flight conditions. The first one is maximum range condition with a velocity of 25 m/s. The second one is maximum endurance condition with velocity 17.5 m/s. Velocities used for the two optimum flight conditions are slightly different from what is listed in table 1 and comes from re-evaluation in detail design phase. In doing CFD simulation for this study, we use full-glider meshing and employing k-ε turbulence model. Objective of this study is to do CFD simulation in the two optimum flight conditions and doing comparison with preliminary design result and previous CFD work in Reference [5].

## 2. Theory

In this study, for the two optimum flight conditions of maximum range with velocity of 25 m/s and maximum endurance with velocity of 17.5 m/s, we deal with a flow with Reynolds number of 1 – 1.5 million. This flow falls into category of high Reynolds number where the inertia forces in the fluid become significant compared to viscous forces. Because in this study we will use an existing code, we should pick from options in the code that suitable to treat that kind of flow. In ANSYS CFX, the code that we employ in this study, we choose to use statistical turbulence model with the use of k-ε turbulence model. It is a "two equations turbulence model" which has advantage of good compromise between accuracy and numerical effort. In this type of turbulence models, velocity and length scale are treated by using separate transport equations (hence the term "two-equation"). In this section we briefly cover about theory that underlying k-ε turbulence models which taken from reference [6] and reference [7].

Statistical turbulence model is based on principle of modifying original unsteady Navier-Stokes equation into average and fluctuating quantities to produce RANS (Reynolds Average Navier-Stokes) equation. However, averaging process introduces additional unknown terms containing products of fluctuating quantities, which acts like additional stresses in fluid. This stress is called "Reynolds" or "turbulent" stress, which difficult to determine and become a new unknown. Reynolds stress should be modelled by additional equation with known quantities so that the equation could reach "closure". The equations used to close the system of equation determine the type of turbulence models. For reaching closure, k-ε turbulence model uses the gradient diffusion hypothesis to relate the Reynolds stresses to the mean velocity gradients and the turbulent viscosity. The k-ε turbulence model relates turbulent viscosity to turbulent kinetic energy and dissipation rate. In this study, we use default k-ε turbulence model provided by ANSYS-CFX solver.

## 3. Methodology

This study is a continuation of previous work of reference [5] for maximum range flight condition only which results in maximum aerodynamics efficiency of 19 for half-glider meshing and 16 for full glider meshing. In reference [5], a benchmarking with similar gliders has been done and validate that the prediction in $C_L$ has been good. Because there is still a discrepancy with preliminary design result of aerodynamics efficiency of 24 or 30, we continue to improve CFD simulation in current study by increasing the amount of grid used from 6 million to 11 million to get an improvement in $C_D$ prediction. Methodology of this study is as shown in figure 1 and will be explained in this section.

Step 1 in figure 1 is to provide geometry of glider GL-1 in IGES from CATIA or SolidWork. Figure 2 shows the geometry we use in this study. Step 2 in figure 1 is doing geometry repair so that the IGES file will be read without error in ANSYS ICEM. The amount of grid generated in step 3 of figure 1 is about 11 million and we use a default unstructured grid with prism layer with $y^+$ equal 1. Step 4 in figure 1 is employing ANSYS-CFX as solver for CFD analysis. Governing equation is RANS with k-ε turbulence model as explained in section 2. Free stream velocity is 25 m/s for maximum range flight condition and 17.5 m/s for maximum endurance flight condition. Computational domain is a square box with boundary condition as follows: inlet at front, outlet at back, and pressure far-field at left, right, top, and bottom. Convergence criterion to achieve is $10^{-4}$. Numerical simulations for both optimum flight conditions are carried out at angle of attack -4 to 12 degree with an increment of 2 degree. Results of step 4 and 5 of figure 1 will be discussed in section 4.

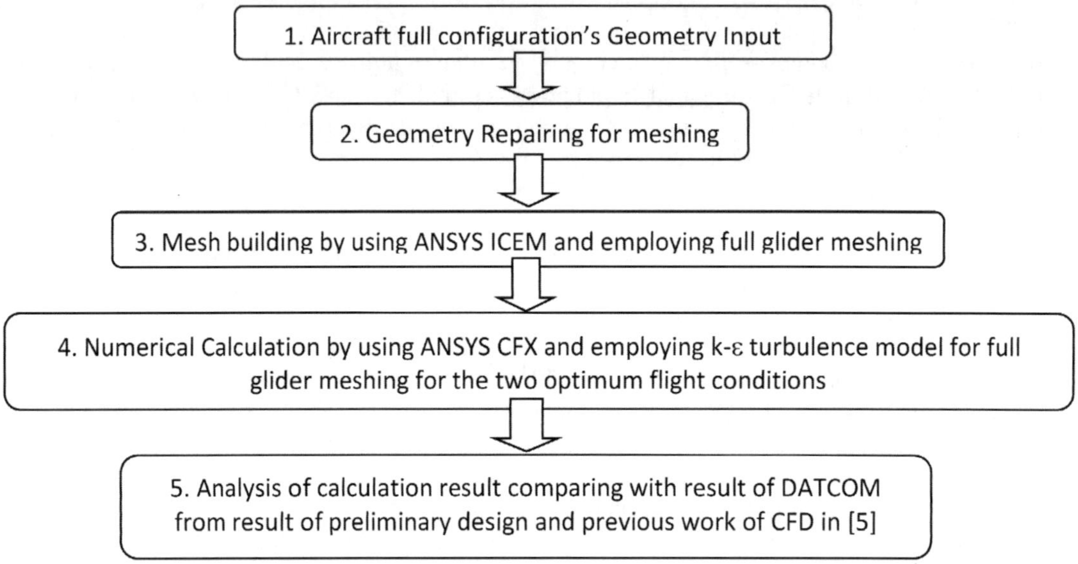


**Figure 1.** Methodology that used in this study

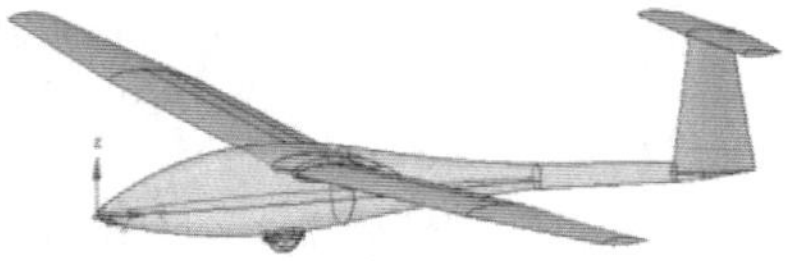

**Figure 2.** Geometry of GL-1 in IGES

## 4. Result and Discussion

In this section we present results of current study or step 4 and 5 in figure 1. In step 4, we do CFD simulation for the two optimum flight conditions, then in step 5 we compare result of step 4 with preliminary design result as in [1], [3] and [4] and previous CFD work of reference [5]. Firstly, we discuss about result for the first optimum flight condition, which is maximum range condition. For this maximum range condition, figure 3 shows result and comparisons for lift coefficient, while figure 4 and 5 show result and comparisons for drag polar and aerodynamic efficiency respectively. We can see from figure 3 that result of current study shows improvement from CFD work of reference [5] of full glider meshing for lift coefficient prediction which has been benchmarked by similar gliders performance.

It means that glider GL-1 has better lift coefficient compared with similar gliders. DATCOM prediction in reference [1] is optimistic for lift coefficient prediction and it is usual according to reference [8] and [9]. While from figure 4 we get that prediction of $C_D$ has been improved compared to previous CFD work of reference [5] for half-glider and full glider meshing, but $C_D$ value is still bigger for the same $C_L$ compared with preliminary design result of DATCOM in reference [1] and reversed engineering result in reference [3] and [4]. The difficulty to get a precise result of $C_D$ is common in CFD. So, further study will be tried to make further improvement in $C_D$ prediction by using better treatment of transition. Figure 5 shows that maximum aerodynamics efficiency from this study is 21 at $C_L$ of 0.91 which is above maximum aerodynamics efficiency of 19 for half-glider and 16 for full-glider in reference [5], but below 24 or 30 that comes from preliminary design phase as listed in table 1. This result comes from the result of $C_D$ prediction that still needs further improvement.

Secondly, we will discuss result of maximum endurance condition. Figure 6 shows result of CFD of climb factor, $(C_L{}^3/C_D{}^2)$, for velocity of 17.5 m/s in this second optimum flight condition. Minimum rate of descent will be obtained when the climb factor, $(C_L{}^3/C_D{}^2)$, is the maximum. In figure 6, it could be seen that it is happened at angle of attack of 4 degree with $C_L$ of 1.08 and the value of climb factor of 418. Now we can calculate rate of descent for gliding angle of 2 degree by using method from reference [2]. We get minimum rate of descent of 0.87 m/s. If we compare with preliminary design which are 0.76 and 0.66 m/s as appeared in table 1, the result of current study is bigger.

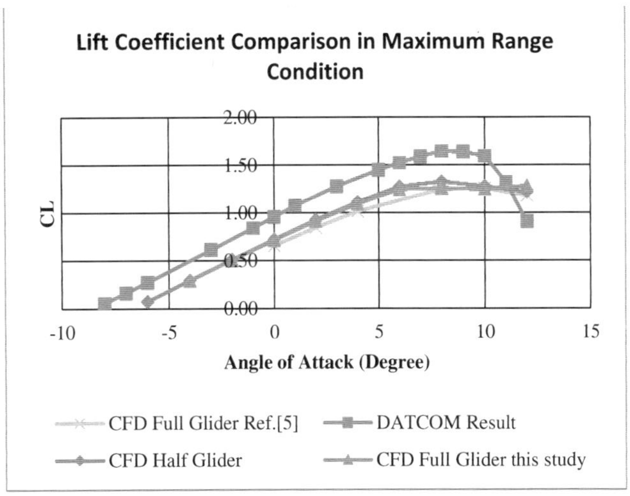


**Figure 3.** Comparison of lift coefficient in maximum range condition

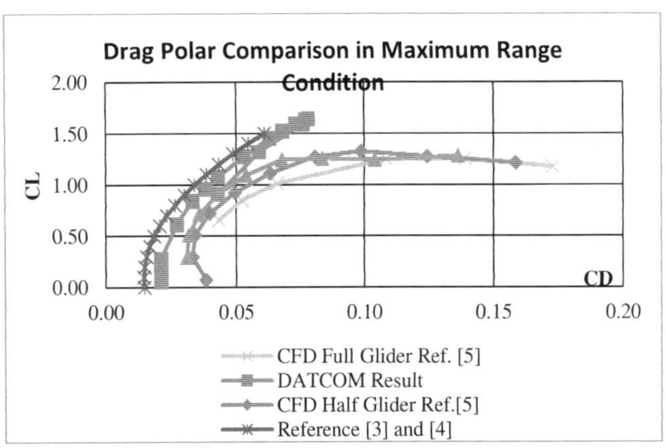


**Figure 4.** Comparison of drag polar in maximum range condition

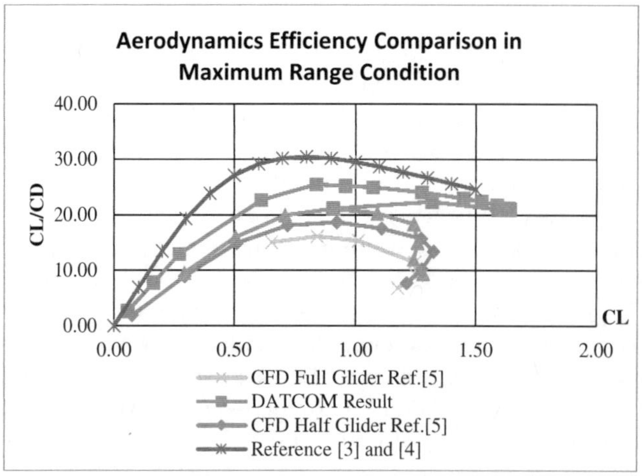


**Figure 5.** Comparison of aerodynamics efficiency in maximum range condition

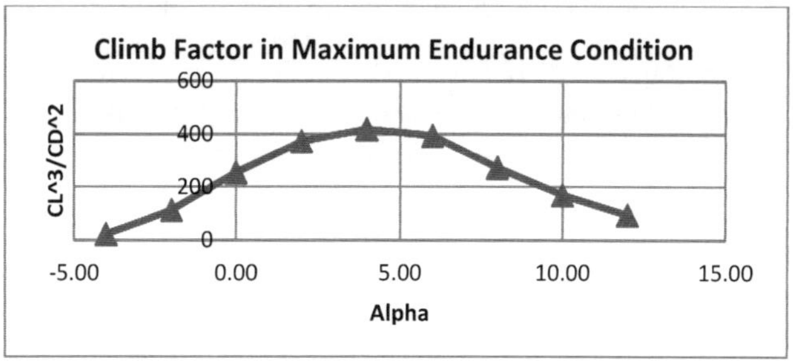


**Figure 6.** Climb factor in maximum endurance condition

Finally, we summarize comparison of aerodynamic performance of glider GL-1 of this study with preliminary design result of reference [1], [3], and [4] in table 1. To calculate range and time, we use the same method as applied to preliminary design result as explained in section 1. From table 1 we can see that, for maximum range flight condition, because of lower aerodynamic efficiency, current study gives smaller maximum range of 9.6 km compared to preliminary design result (11 km and 13.7 km). As for maximum endurance flight condition, because of bigger minimum rate of descent, we obtain smaller maximum endurance from current study of 13.1 minutes compared to preliminary design result (14.4 minutes and 15.9 minutes). The root cause of this result is $C_D$ prediction by CFD that is bigger compared to preliminary result by using DATCOM [1] and by using reversed engineering method [3-4]. Therefore, improvement in treating transition will be tried in a near future to improve accuracy of drag prediction or by using another meshing scheme as presented by other [10].

**Table 1.** Performance comparison of GL-1 of current study and preliminary design result

| | With Data of Reference [1] | With Data of Reference [3] and [4] | Current study |
|---|---|---|---|
| **In Maximum Range Condition:** | | | |
| (CL/CD)max | 24 | 30 | 21 |
| CL | 1.18 | 0.8 | 0.91 |
| CD | 0.048 | 0.027 | 0.043 |
| V (m/s) | 24.3 | 22 | 25 |
| t (minute) | 7.5 | 10.4 | 6.4 |
| Maximum Range (km) | 11.0 | 13.7 | 9.6 |

| | | | |
|---|---|---|---|
| **In Maximum Endurance Condition:** | | | |
| (RD)min (m/s) | 0.76 | 0.66 | 0.87 |
| V (m/s) | 18.5 | 17.5 | 17.5 |
| Vaverage (m/s) | 12.9 | 12.7 | 11.7 |
| Range (km) | 11.1 | 12.1 | 9.2 |
| Maximum Endurance (minute) | 14.4 | 15.9 | 13.1 |

## 5. Conclusion

CFD simulation done in this study for maximum range flight condition gives result of maximum aerodynamic efficiency of 21 at $C_L$ of 0.91 that lower from results of preliminary design phase which are 24 at $C_L$ of 1.18 or 30 at $C_L$ of 0.8, by using different methods. However, results of this study for the first optimum flight condition is better than our previous CFD work, with the value of aerodynamics efficiency of 21 compared with 19 from half-glider meshing and 16 from full-glider meshing. As for maximum endurance flight condition, CFD simulation study of this study, gives minimum rate of descent of 0.87 m/s that bigger from results of preliminary design phase of 0.76 m/s from other work, and 0.66 m/s using our current method. Moreover, by using national competition rule and method, we get maximum range of 9.6 km compared with 11 km or 13.7 km of preliminary design result. Also, with the same method, we get maximum endurance of 13.1 minute compared with 14.4 minute or 15.9 minute. The root cause of this result mentioned above is prediction of $C_D$ that still need further improvement.

**Acknowledgments**
This study is funded by P3MI research funding of Institute of Technology Bandung. We wish to thank to Dr. Taufiq Mulyanto and team of design group for providing data of glider GL-1.

**References**

[1] Pratama H A 2015 *Conceptual Design of Indonesian National Glider GL-01* (Bandung: Institute of Technology Bandung).
[2] Ruijgrok, G J J 1990 *Elements of Airplane Performance* (Delft: Delftse Universitaire Pers) p 306-329.
[3] Kurniasari N A D 2016 *Estimation of Aerodynamics Parameter of National Glider GL-01 By Reversed Engineering (In Bahasa Indonesia)* (Bandung: Institute of Technology Bandung).
[4] Kurniasari N A D, Nurhakim M L I, Arifianto O and Mulyanto T 2016 Glider Uncentered Drag Polar Estimation Based on Flight Manual Data *Advance in Aerospace Science and Technology in Indonesia vol 1* p 118-125.
[5] Amalia E A, Moelyadi M A, Julistina R, Putra C A 2018 Aerodynamics Characteristics of Glider GL-1 based on Computational Fluid Dynamics *International Seminar on Aerospace Science and Technology VI/2018, 25 – 26 September 2018.*
[6] ANSYS Inc. 2009 *Ansys CFX-Solver Theory Guide* (Canonsburg PA: ResearchGate) Release 12.1 p 53-61.
[7] Tenneke H and Lumley J L 1972 *A First Course in Turbulence* (Massachusets USA: MIT Press).
[8] Olson E D 2015 Semi-Empirical Prediction of Aircraft Low-Speed Aerodynamic Characteristics *AIAA Science and Technology Forum (SciTech 2015)* https://ntrs.nasa.gov/search.jsp?R=20150006019 p 10.
[9] Rithice R W 2013 *Comparative Evaluation of Computational Techniques for Estimating UAV Aerodynamic, Stability and Control Derivatives* (Minnesota: University of Minnesota) https://conservancy.umn.edu/bitstream/handle/11299/157177/Ritchie_umn_0130M_13734.pdf p 40-43.
[10] Boermans L M M 2006 *Research on Sailplane Aerodynamics at Delft University of Technology. Recent and Present Developments* (Delft: TU Delft) p 9-17.

IOP Conf. Series: Materials Science and Engineering **645** (2019) 012007 doi:10.1088/1757-899X/645/1/012007

# Evaluation and Improvement of Unscheduled Removal Component Reliability Calculation Method with No Fault Found (NFF) Shop Finding Results

**F K Suciati and E Suwondo**

Aerospace Engineering, Bandung Institute of Technology, Indonesia

Email: esuwondo@gmail.com

**Abstract**. The test results of the removed aircraft components are classified into No Fault Found (NFF), Real Reason Conform (RRC) and Other Defect Found (ODF). Corrective actions will be taken on findings that are classified as RRC and ODF, while for the NFF category no action will be taken. In the MRO XYZ, for the NFF category, the age of the component during the removal is included into the failure data. Moreover, the reliability calculation in the MRO XYZ uses the 2-parameters Weibull distribution. These will result in unrepresentative reliability calculation. Therefore, it is necessary to evaluate the method. The research compares the results of the reliability calculation between the 2-parameters and the 3-parameters Weibull distributions by applying them to several part numbers data. Furthermore, the reliability calculation using the 3-parameters Weibull is also applied to the data where the NFF category is not included the failure data. The location parameter, $\gamma$, is calculated using the Muralidhar method. The research also develops an excel VBA macro-based tool for reliability calculation. The results of the reliability calculation show that there is a significant difference of the failure rate between using the 2-parameters and the 3-parameters Weibull distribution. The failure rate calculated by using the 3-parameters Weibull distribution for the data where NFF category is not included into the failure data show a significant different as compared to the method used by the MRO XYZ, where the difference is affected by the percentage of NFF items in the failure data.

## 1. Introduction

In carrying out aircraft maintenance, components are usually removed. Component removal are grouped into two, namely scheduled and unscheduled removal. After the component is removed, then they are tested in the store to find the causes (shop finding) and then it can be determined whether the component needs repair or not.

Shop finding are categorized into 3 types namely Real Reason Conform (RRC), Other Defect Found (ODF) and No Fault Found (NFF). RRC is called if the results of shop finding are the same as the reason of removal reported by pilots or technicians on line maintenance. ODF is called if the results of shop finding are different from the reason of removal reported by pilots or technicians on line maintenance. Whereas NFF is called if the testing in the shop turns out that there is no failure on the component.

The NFF will be a problem if it occurs on components with unscheduled removal. If the pilot or technician reports that there has been a failure in a component so that the component needs to be removed while it turns out that shop finding states that the component is fine, then this component needs further attention. However, what has happened so far is if there is a NFF result on a component with an unscheduled removal, without any improvement, the components are immediately installed again on the aircraft. This causes the component being removed frequently with the same shop finding results, namely the NFF so that the component can be called low reliability. In addition there is also an inappropriate method in calculating component reliability. Therefore, it is necessary to improve the reliability calculation method before making improvements to the NFF component handling method at the shop.

The purpose of this research is to evaluate the reliability calculation method of NFF items used and to design a new method to improve the calculation of reliability of NFF items.

The methodology used, first is formulating the research background, then formulating the problem from the results of the background description that has been carried out together with the literature review, then defining the research boundaries. After that, evaluating the methods used in calculating the reliability of components that experienced a lot of unscheduled removal with shop finding NFF results, designing new methods to improve the calculation of component reliability that experienced unscheduled removal with shop finding NFF results, designing data formats and application procedures, testing new methods for some samples of items that have been designed into a data format in such a way, analyze the results of reliability calculations between the methods currently used (existing) and new methods.

## 2. Literature Review

Maintenance is a combination of various actions taken to safeguard an item or repair it until an acceptable condition. In the maintenance of aircraft components, removal is usually carried out. There are two types of removal namely scheduled removal and unscheduled removal. Scheduled removal is a removal that is done according to a predetermined time interval. Scheduled removal is usually done if there are order documents such as AMI, EO and TO.

In this research, it will be limited to the calculation of weibull distribution parameters using parametric analysis used to determine failure rate [1].

In this research, Weibull distribution is expressed in 3 parameters. The three parameters are β (shape parameter), η (scale parameter) and γ (location parameter). Shape parameter, β can be used to state the failure condition as time goes by. Value of the shape parameter $\beta < 1$ (infant mortality) explains the very high failure rate at the beginning of age (early death) and the action to be taken is system modification. Value of the shape parameter $\beta = 1$ (random failures) describes a failure rate that is constant with age as failure does not depend on time. The action that must be taken is the modification of the component. Whereas the value of the $\beta > 1$ form parameter (wear out failures) explains that the failure rate will increase with age and the action to be taken is inspection or restoration [2]. To prove whether the data really has a Weibull distribution type distribution or not, the Mann test is used. The formula for the statistical test is as follows:

$$M = \frac{k_1 \sum_{k_1+1}^{r-1}[(\ln t_{i+1} - \ln t_i)/M_i]}{k_2 \sum_{i=1}^{k1}[(\ln t_{i+1} - \ln t_i)/M_i]} \quad (1)$$

Where $k_1 = \left\lfloor \frac{r}{2} \right\rfloor$, $k_2 = \left\lfloor \frac{r-1}{2} \right\rfloor$, $M_i = Z_{i+1} - Z_i$ and $Z_i = \ln\left[-\ln\left(1 - \frac{i-0.5}{n+2.5}\right)\right]$. As for r is the total number of data, n is the number of failures that occur, $k_1$ is the amount of data before the median, $k_2$ is the amount of data after the median, $2k_1$ is denumerator and $2k_2$ is the numerator. If the value of $M > F_{crit}$ then the failure time is not Weibull. The $F_{crit}$ value is the value of the function of the numerator and denumerator using the statistical function in Excel that is F.INV.RT with a value of $\alpha = 5\%$.

The failure rate of the Weibull λ (T) distribution can be stated as follows:

$$\lambda(T) = \frac{\beta}{\eta} \left(\frac{T-\gamma}{\eta}\right)^{\beta-1} \quad (2)$$

For the shape parameter above 1 the failure rate will increase. If the shape parameter is equal to 1, the failure rate tends to be constant. Whereas for the shape parameter value is less than 1, the failure rate tends to decrease.

In this research the method used to calculate Weibull parameters is least square method along with probability plotting methods. The least square method is a standard approach in regression analysis to approach the solution of overdetermined systems. The least square method is also used in calculating the index of fit of the data. The index of fit value states that the data level follows the Weibull distribution. If it gets closer to 1 then the data gets better and more closely follows the Weibull distribution. Probability plotting in this reserach uses Excel to reduce the factor of high subjectivity. The steps to determine the Weibull distribution parameters can be derived from the following reliability equation [1]:

$$R(T) + Q(T) = 1 \tag{3}$$

where R (T) is the Weibull reliability function and Q (T) is a Weibull unreliability function. So that we will get the following equation:

$$1 - Q(T) = e^{-(\frac{T-\gamma}{\eta})^{\beta}} \tag{4}$$

By reversing the two terms, the following equation is obtained:

$$\frac{1}{1-Q(T)} = e^{(\frac{T-\gamma}{\eta})^{\beta}} \tag{5}$$

By applying a natural logarithm operation, the following equation is obtained:

$$ln\frac{1}{1-Q(T)} = (\frac{T-\gamma}{\eta})^{\beta} \tag{6}$$

Then the natural logarithm operation is applied again, then the following equation is obtained:

$$ln\,ln\frac{1}{1-Q(T)} = \beta\, ln(T-\gamma) - \beta\, ln\,\eta \tag{7}$$

The equation above is an equation of a straight line Y = mX + C with each of them are:

$$Y = ln\,ln\frac{1}{1-Q(T)} \tag{8}$$

$$m = \beta \tag{9}$$

$$X = ln(T-\gamma) \tag{10}$$

$$C = -\beta\, ln\,\eta \tag{11}$$

The probability plotting method with uncensored data is first done, which is to sort the age to failure (Time to Failure) and give a data index of j = 1 as the time data to 1 to N where N is the number of data failures. Furthermore, it is calculated the Median Rank value for data to j and the number of sample N data with the equation [2]:

$$MR = Q(T_j) \times 100 \approx \frac{j-0.3}{N+0.4} \times 100 \tag{12}$$

Median Rank is the value of a probability of failure occurring in the data to j from the sample as much as N. In censored data, the index j in determining the median rank will not be the same as the value of N. For probability plotting and least-square methods in the data censored value of N in the calculation the median rank is the value of the number of samples. So that a table can be made to simplify calculations by using probability plotting methods with censored data. Calculation of location parameters γ uses Muralidhar method because there is no subjectivity factor involved in the calculation. Assuming that the time to failure (Time to Failure) that has been recorded has been sorted from the smallest to the largest, the equation for calculating location parameters is [1]:

$$\gamma = \frac{T1Tn-T_j{}^2}{T1+Tn-2Tj} \tag{13}$$

The index j becomes $j_{censored}$, namely $j_{censored} = [np(n/N)]$. While the p value is as follows:

$$p = 0.8829n^{-0.3437} \tag{14}$$

The reliability calculation methods currently used and their improvements are as follows:

(a) using the 2-parameter Weibull distribution (the location parameter nol is zero) is fixed to a 3-parameter Weibull distribution,

IOP Conf. Series: Materials Science and Engineering **645** (2019) 012007 doi:10.1088/1757-899X/645/1/012007

(b) removal with shop finding NFF results that are considered as a repaired failure to be not included in the failure data,
(c) TSI values less than 1000 FH are considered as repaired data outliers to remain included in the inserted to analysis category.

To calculate the reliability of a component, failure data is grouped according to part number. Then after that it is classified again based on its failure mode, because the reliability calculation is specific to one failure mode. The TSI value obtained cannot be directly entered into the reliability calculation. TSI values are arranged based on the results of shop finding. If the shop finding results show that the component (in this context part number) has failed, then the component's age is calculated from the start installed on the plane. However, if shop finding of these components shows the results of the NFF, then the age of the component has not stopped when removing, the TSI still continues when installed on the plane (not starting from zero again).

## 3. Improvement and Application of The New Method

In this research the authors developed Excel-based tools for reliability calculations that are easy to use. The tool used is based on Excel VBA macros. By using VBA macros, the user only needs to enter the failure data and press the run button, then the desired output will be automatically obtained. The script in the VBA macro contains the Weibull functions used in this study. Then, the user needs to enter failure data in the 'i' and 'no' columns. After entering the data, then the user needs to press the run button. After that, the Mann test results will appear whether the data is Weibull distributed or not. If not, the "data does not fit with weibull distribution" dialog box will appear, but if we have a Weibull distribution, a dialog box will appear to enter the number of samples [4].

**Figure 1**. Excel based tools for reliability calculations

The next step is to use the 'Data Analysis' tool to get the values of the shape parameter β and the scale parameter η. After selecting the 'Data Analysis' tool, then choose the 'Regression' tool in the 'Analysis Tools' option. In the 'Regression' tool box dialog above for the option 'Input Y Range', enter the ln value (ln (1 / (1-Median Rank))) from the first row (column name) to the last line containing the failure data. While for the option 'Input X Range', enter the value ln (ti) from the first row (column name) until the last line containing the failure data. After that, add a check mark in the 'Labels' box. Next to the option 'Output options', select 'New Worksheet Ply' so that the calculation results appear on a sheet that is different from the data. Finally on the 'Residuals' option, select 'Line Fit Plots' then click OK. Make excel will display the regression results and the results will appear in the new sheet. Scale parameter value η = e ^ (- intercept / β). Furthermore, if the value of the shape parameter β and the scale parameter η is obtained, what needs to be done is to make a plot of the failure rate.

In this research, the comparison of existing methods with new methods was applied in 4 case studies namely part numbers 071-01503-2601, 398908-3, 8055515 and 822-0990-004. From the Mann test results, all three were proven to have Weibull distribution [4-5].

### *3.1. PN 071-01503-2601*

In the 2 parameter Weibull distribution, the existing method produces the shape parameter value β = 1,557 hr and the scale parameter η = 5972,836 hr. While the new method produces the shape parameter value β = 0.763 hr and the scale parameter η = 10748,973 hr. In the 3-parameter Weibull distribution, the existing method produces the shape parameter value β = 0.993 hr, the scale parameter η = 4922.18 hr and the location parameter γ ≈ 864 hr. While the new method produces the shape parameter value β = 0.52 hr, the scale parameter η = 11824,498 hr and the location parameter γ ≈ 147 hr.

### *3.2. PN 398908-3*

In the 2 parameter Weibull distribution, the existing method produces shape parameter values β = 1,597 hr and scale parameters η = 6570,178 hr. While the new method produces the shape parameter value β = 0.831 hr and the scale parameter η = 4454.63 hr. In the 3-parameter Weibull distribution, the existing method produces the shape arameter value β = 0.835 hr, the scale parameter η = 5375.41 hr and the location parameter γ ≈ 1027 hr. While the new method produces the shape parameter value β = 0.725 hr, the scale parameter η = 4542,839 hr and the location parameter γ ≈ 41 hr.

### *3.3. PN 8055515*

In the 2 parameter Weibull distribution, the existing method produces the shape parameter value β = 2,274 hr and the scale parameter η = 7934,89 hr. While the new method produces the shape parameter value β = 0.603 hr and the scale parameter η = 6005.25 hr. In the 3 parameter Weibull distribution, the existing method produces the shape parameter value β = 0.998 hr, the scale parameter η = 6221,620 hr and the location parameter γ ≈ 1191 hr. While the new method produces the shape parameter value β = 0.383 hr, the scale parameter η = 7373.904 hr and the location parameter γ ≈ 25 hr.

### *3.4. PN 822-0990-004*

In the 2-parameter Weibull distribution, the existing method produces the shape parameter value β = 1,799 hr and the scale parameter η = 9354,004 hr. While the new method produces the shape parameter value β = 0,522 hr and the scale parameter η = 8254,707 hr. In the 3-parameter Weibull distribution, the existing method produces the shape parameter value β = 1.414 hr, the scale parameter η = 1.414 hr and the location parameter γ ≈ 786 hr. While the new method produces the shape parameter value β = 0.573 hr, the scale parameter η = 8212.335 hr and the location parameter γ ≈ -23 hr.

## 4. Analysis of Calculation Results

### *4.1. PN 071-01503-2601*

Comparison of failure rate graphs for PN 071-01503-2601 between the existing method and the new method for 2-parameter Weibull distribution and for the 3-parameter Weibull distribution, as follow:

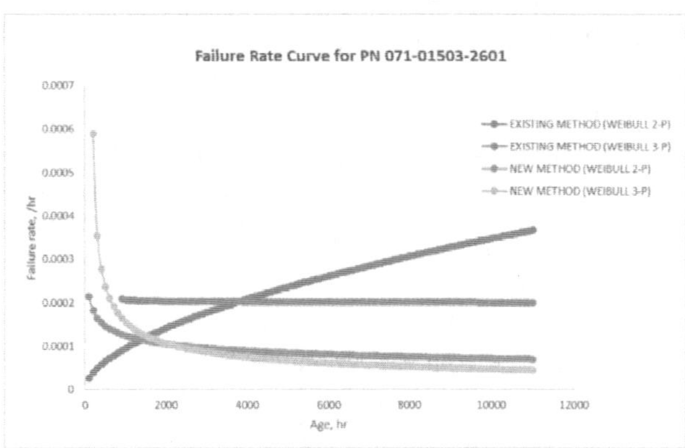


**Figure 2.** Failure rate curve for PN 071-01503-2601

*4.2. PN 398908-3*

Comparison of failure rate graphs for PN 398908-3 between the existing method and the new method for 2-parameter Weibull distribution and for the 3-parameter Weibull distribution, as follow:

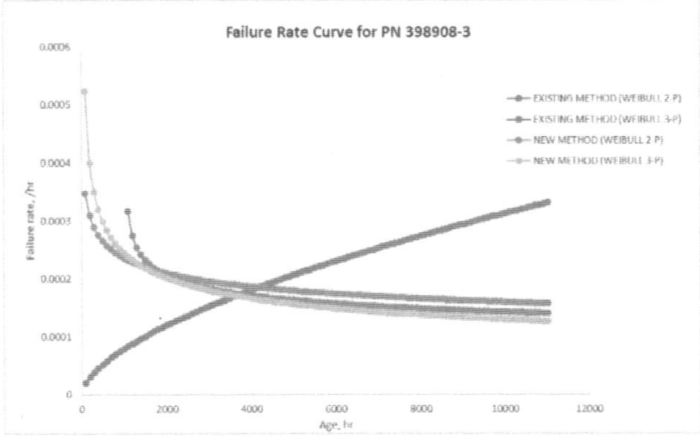


**Figure 3**. Failure rate curve for PN 398908-3

*4.3. PN 8055515*

Comparison of failure rate graphs for PN 8055515 between the existing method and the new method for 2-parameter Weibull distribution and for the 3-parameter Weibull distribution [6], as follow:

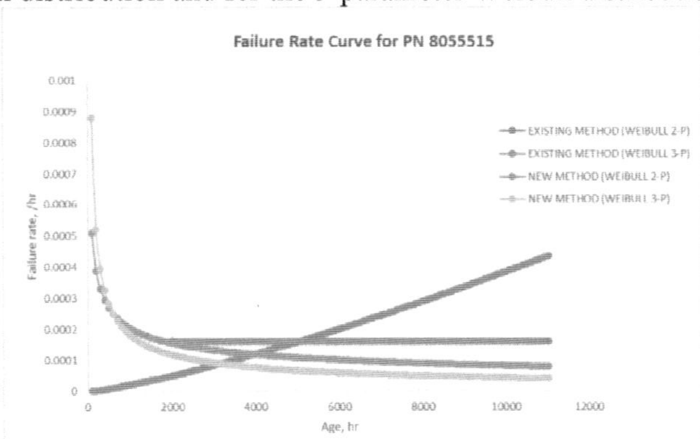


**Figure 4**. Failure rate curve for PN 8055515

*4.4. PN 822-0990-004*

Comparison of failure rate graphs for PN 822-0990-004 between the existing method and the new method for 2-parameter Weibull distribution and for the 3-parameter Weibull distribution, as follow:

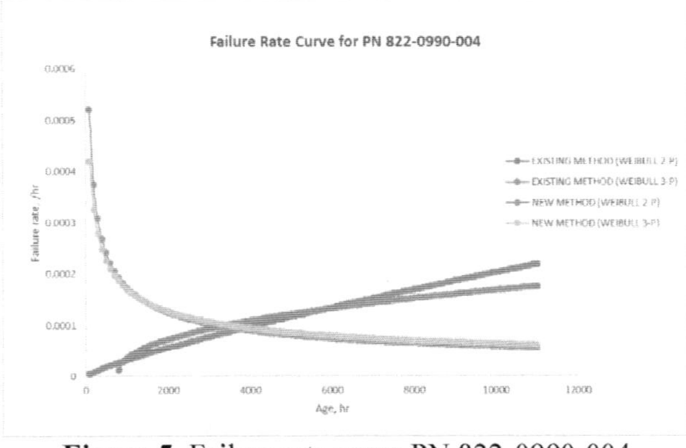


**Figure 5**. Failure rate curve PN 822-0990-004

The percentage of NFF for each part number is shown in the graph below.

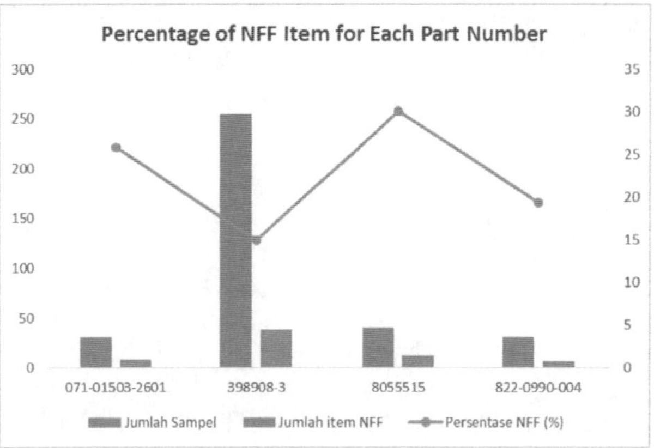


**Figure 6**. Percentage of NFF Item for each part number

## 5. Conclusions

The method used which considers the NFF as a failure cannot be justified should the NFF not be considered a failure. The new method has a more consistent form of failure rate curve between the 2 and the 3 parameter Weibull. The results of the reliability calculation show the difference in the parameter value of the form β which depends on the percentage number of NFF in the data sample. The use of 2-parameter Weibull in reliability calculations cannot be justified, it should use a 3-parameter Weibull calculation. The reliability calculations should use a 3-parameter Weibull distribution so that component failures before operation can also be detected. Making user-friendly Excel calculations using VBA macros can be implemented and searching for the life of an item needs to be done in more detail from the beginning of the item being put into operation.

## References

[1] C. E. Ebeling, An Introduction to Reliability and Maintainability Engineering, New York: The McGraw-Hill Companies Inc., 1997.

[2] D. Kececioglu, Reliability Engineering Handbook Volume 1, Arizona: Prentice-Hall Inc., 1991.

[3] R. B. Abernethy, The New Weibull Handbook, Florida: Robert B. Abernethy, 2006.

[4] R. B. Abernethy, The New Weibull Handbook, Florida, 2006.

[5] Dorner, William W, "Using Microsoft Excel for Weibull Analysis', Quality Digest, October 1997.

[6] www.reliasoft.com/Weibull/features1.htm. accessed on June, 4th, 2018.

# Developing new aviation management postgraduate program in responding to Industry 4.0: Key findings from multistages-multilevels market study

**N H Ghafar[1,4], N A A Rahman[1], M F N Mohammad[1], M Z Shah[2] and F Hassan[3]**

[1] Universiti Kuala Lumpur, Malaysian Institute of Aviation Technology, Persiaran A, Off Jalan Lapangan Terbang Subang, 47200 Subang, Selangor, Malaysia
[2] Centre for Innovative Planning & Development, Faculty of Built Environment & Surveying, Universiti Teknologi Malaysia, 81310 Skudai, Johor, Malaysia
[3] Faculty of Business & Management, Universiti Teknologi MARA, 40450 Shah Alam, Selangor, Malaysia
[4] Putra Business School, Universiti Putra Malaysia, 43400 Serdang, Selangor, Malaysia

Corresponding email: hisyam@unikl.edu.my

**Abstract**. Most aviation related academic programs in Malaysia were designed centred around the aerospace engineering and aircraft engineering technologies. However, the emergence of Industry 4.0 demands aviation talents especially the non-management aviation practitioners such as aircraft engineers and other technical personnel to equip themselves with certain management-related skills. Hence, there is a pressing need for an advanced industry-driven and practitioners-based academic program in aviation management suits for various categories of aviation practitioners. The aims of this study are to explore the employability prospective of aviation managers (first stage); to examine the Industry 4.0-related management skills needed by aviation companies (second stage); and to analyse the personal development intention of individual practitioners (third stage). This study employed multistage-multilevel analysis on a single-case market study in a Malaysian aviation technical training university. Findings from this study are segregated into three levels. At the industry level, it is concluded that there is a brighter employability prospect of aviation managers at least until 2030. At the organizational level, there is an urgent need for the aviation professionals to upgrade themselves with managerial skills relevant to the Industry 4.0 such as data driven scenario planning, big data analysis, and collaborative decision making. At the individual practitioner's level, finding indicates that there is an increasing awareness among the individual aviation professional in particular the technical and engineering professional to engage in personal development in aviation management at postgraduate level. The key contribution of this case study is that, through multistages and multilevels analysis, it not only expands the robustness of the market study on new postgraduate program development but also able to incorporate the perspectives of different aviation stakeholders group i.e. the aviation industry, aviation organizations, and individual aviation talents. This multistage-multilevel model can facilitate university's policy makers to design future industry-based and practitioners-driven development program meant for industry practitioners.

## 1. Introduction

Aviation industry is one of the most important industries for Malaysia. It generates high value, high technology, and high income jobs [1], which is among the key element to achieve the status of developed nation by year 2020 [2]. This is why the government of Malaysia put greater emphasis and investment on the development of aviation talent through education and training. At this moment, most aviation related academic programs in Malaysia are purely centred around the aerospace engineering and aircraft engineering technologies. This is due to the fact that the vast components of Malaysian aviation industry represented by aircraft components' manufacturing sub-sector as well as maintenance, repair, & overhaul (MRO) sub-sector [3].

However, the introduction of new-generation aircraft especially by the duopoly aircraft manufacturers i.e. Airbus and Boeing adopting latest technology [4] has led to the increasing demand for new advanced skills development of the next generations of aviation human capital. Adding to that, the recent emergence of Industry 4.0 (IR 4.0) demands aviation talents including the non-management practitioners such as aircraft engineers and other technical aviation personnel to equip themselves with certain management-related skills.

Employing the stakeholders' theoretical perspective, this market study aims for four objectives. First, it investigates the future employability trend within the industry. Second, it investigates the emerging Industry 4.0-related managerial skills in aviation. Third, it examines the organizations' employment demand for IR 4.0-related management skills. And fourth, it analyzes the need for management advanced development among individual technical aviation practitioners. By addressing these four objectives, this study intends to create a more realistic and industry-driven basis for developing a postgraduate aviation management program.

Past literature in academic program development has put greater emphasis on providing the skills improvement intervention that can overcome the unemployment problem in the society [5]. The designs of academic program at that time were guided based on two competing schools of thought in educational theory i.e. cognitism and behaviorism. Cognitism approaches the education development more towards intellectual transfer of learning, whereas behaviorism focuses the cultivation of learning through stimulus and behavioral response of the learners [6].

However in recent years, the focus of an academic program has shifted from not just preparing the graduate for a single 'employment', but also retaining the graduates' 'employability' i.e. sustaining the graduates' industry-relevant career throughout their career tenure. The alumni of the academic program are now expected to become the future problem solvers, change agents, and transition managers that suit with the rapid changing nature of the industry [7].

As one of the strategic industry, aviation is known as the area with most rapid changing technology due to the advanced technology used by aircraft manufacturers, airlines operators, and aviation support services. The development of academic training program are mainly guided by internationally recognized aviation regulations such as Part 147 requirements for maintenance training organizations, or Part 145 requirements for maintenance repair organizations.

Nevertheless, the emergence of Industry 4.0 within the aviation industry has extended the focus of aviation academic programs into equipping the supply of aviation talent not only with aviation technical competency, but also with managerial skills involving multiple levels of direct and indirect stakeholders within the aviation industry. Present literature in academic program development especially in aviation education is lacking in providing the guide on how an industry-based and practitioners-driven academic program should be developed in a systematic way.

In order to have an industry- and practitioners-driven academic program that addresses the multiple stakeholders' needs, its development should be systematically done in stages involving multiple parties whose interest vested in it. These stages involve exploring employability at industry level, converging the industry employability with organizational employment at the companies' level, and confirming personal career need at the individual level. Based on that, the following multistages and multilevels market study was conceptualized as per depicted in Figure 1.

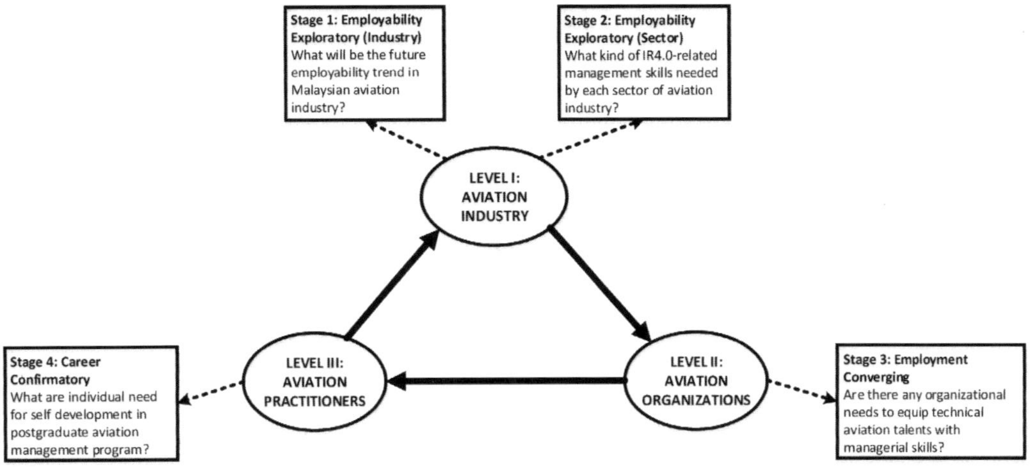


**Figure 1**: Conceptual framework of multistages and multilevels market study

The development of this conceptual framework is theoretically underpinned by the Stakeholders Theory. The Stakeholders Theory [8] stipulates that those 'individual or group who is affected by or can affect the actions or outcomes of an organization are the organization's stakeholder'. Applying this into the market study for aviation management postgraduate program development, the aviation industry, the aviation companies, and aviation practitioners are direct beneficiaries of the academic program. The involvement of industry stakeholders into the curriculum development has been found to be able to enhance the program quality of a higher education institution [9].

## 2. Methodology

The multistage data collection refers as the investigation which involves three or more stages of sequential and converging processes that requires multiple approaches of gathering information from the sources [10]. This study employed the multistage data collection in order to explore certain scenario pattern within the population, to explain the reason, and to confirm the reason [10].

The rationale of collecting data at different stages is due to the initial need to establish the industry employability sustainability first before assessing the employment in the individual organization and individual level consecutively. This pre-requisite consideration is based on the fact that in designing any new academic program, it should firstly contribute to the graduates' employability with the industry [11]. Over past few years, there is a shift in trend among professional workforce to search for the 'security of employability' rather than the 'security of employment' [12]. Which mean that majority of talented individuals intentionally choose not to be loyal to a single organization, but rather loyal to his/her career. Therefore, the 'employability' should become the prerequisite to the 'employment' and 'personal career development' needs.

### *2.1. Stage 1/Level I: Exploring Employability Trend within Aviation Industry*

This market study process started with the exploration of employability trend in Malaysian aviation industry (i.e. Level I). A content analysis data collection approach was used to analyze the number of aviation companies, number of aviation job titles created, number of aviation workforce, and number of managerial positions shortage in the industry. Data obtained from seven Malaysian aviation industry reports published by ministries, government agencies, and independent consultant were extracted and analyzed to establish the statistic on aviation manpower demand until the year 2030.

### *2.2. Stage 2/Level I: Explore the Industry 4.0 Managerial Skills Needs*

Next stage involved the process of exploring categories of emergent Industry 4.0-related managerial skills in each sectors in Malaysian aviation industry (i.e. Level I). Employing thematic analysis

IOP Conf. Series: Materials Science and Engineering **645** (2019) 012008 doi:10.1088/1757-899X/645/1/012008

technique, information from eight published industry reports and eighteen journals articles on Industry 4.0 in aviation were extracted. The information were codified using Atlas.ti and relevant themes were identified to represent different categories of emerging new skills related to the application of Industry 4.0 in every aviation industry value chain.

*2.3. Stage 3/Level II: Converge the Industry Employability with Organizational Employment*

This stage 3 involved data collection at the organizational level (i.e. Level II). The process involved separate qualitative interview with three organizational leaders from two reputable aviation companies and an aviation regulatory agency in Malaysia. The first respondent is the Vice President (Engineering and Maintenance) of an airline, the second respondent is the Chairman of the Board of Directors of a MRO company, and the third respondent is the Principal Assistant Director (Air Traffic Management) of an aviation regulatory agency. Collected interview responses were then recorded, transcribed, and analyzed by three experienced qualitative researchers using Atlas.ti software.

*2.4. Stage 4/Level III: Confirming Career Need for Personal Development*

The last stage involved quantitative survey that was randomly administered online using Google Form to different groups of experienced and fresh aviation practitioners (i.e. Level III). Sampling frame were driven from university-industry networking records and bachelor of aviation management alumni database. One hundred participation invitations were sent, however only 55 responded which made 55 percent response rate. The details of the surveyed respondents are as per Table 1 below. Collected data were analyzed descriptively using SPSS.

Findings from all four stages above were then consolidated in a separate report and being presented, commented, and endorsed by a panel of experts consisted of three persons i.e. two academic experts from two higher education institutions and one expert from the aviation industry.

**Table 1**: Demographic profile of the surveyed respondents

| Category | Sub-categories | Quantity | % |
|---|---|---|---|
| Organization types | MRO | 8 | 15 |
| | Operation | 14 | 25 |
| | Production | 3 | 5 |
| | Aviation Training | 16 | 29 |
| | Continuous Airworthiness | 1 | 2 |
| | Police Air Wing | 1 | 2 |
| | Business | 4 | 7 |
| | Customer Services | 1 | 2 |
| | Design & Certification | 2 | 4 |
| | IT | 2 | 4 |
| | Logistic | 3 | 5 |
| Job category | Managers | 9 | 16 |
| | Executives | 12 | 22 |
| | Engineers | 4 | 7 |
| | Pilot | 1 | 2 |
| | Instructors | 4 | 7 |
| | Officers | 12 | 22 |
| | Management Trainees | 1 | 2 |
| | Alumni & Students | 11 | 20 |
| | Not specified | 1 | 2 |
| ***Total Respondents*** | | ***55*** | ***100*** |

## 3. Results

### *3.1. Employability Trend in Malaysian Aviation Industry*
The study reveals that there is a brighter employability prospect in Malaysian aviation industry until the year 2030. The industry witnessed a tremendous increase of the number of aviation companies i.e. from 50 (1998) to 266 (2017) companies. On top of that, the Malaysian aviation industry alone contributes to the creation of 483 aviation specific job titles. Concurrent to the increase of the number of aviation companies, the number of aviation workforce is also shown an upsurge i.e. from 6,800 (1998) to 21,200 (2017), and expected to increase further to 32,000 by 2030. Similarly, the aviation managerial positions had increased from 680 (1998) to 2,120 (2017) and anticipated to increase further to 3,200 by 2030.

### *3.2. Emerging Industry 4.0-Related Managerial Skills in Malaysian Aviation Industry*
The study finds several emerging managerial skills that are related to the Industry 4.0 throughout the aviation industry value chain. Among identified Industry 4.0-related managerial skills are analytical thinking and decision making, innovative thinking, collaborative planning and decision making, scenario planning and visualization, big data management, and real-time data driven decision making.

### *3.3. Employment needs of management competencies for technical aviation practitioners*
The interviews data with aviation companies' top leaders also shows that due to the cost factors, there is an urgent need for engineering and technical aviation practitioners to add on some management skills into their existing engineering and technical skills.

### *3.4. Individual practitioners' needs for academic development in aviation management at postgraduate level*
The survey result indicates strong urge for advanced self-development among aviation practitioners. 93 percent of the respondents agreed that obtaining higher academic qualification significantly assist their vertical as well as horizontal career movement. In addition, the urge for personal development at postgraduate level are stronger among experienced aviation practitioners. The main reason behind this is that experienced practitioners realize that continuing education is also part of organizational needs.

Table 2 below summarizes in detailed the key findings of this market study at each stage and level.

**Table 2**: Study stages, data collection method, sources of data, and key findings

| **Stage *(Level)*/Objectives** | **Data Collection Method** | **Sources of data & Sample** | **Key Findings** |
|---|---|---|---|
| **Stage 1: Exploratory Stage [Prerequisite]** *(Industry level)* To explore the employability prospect of aviation management postgraduate. | Content analysis | Aviation industry reports ($n_{reports}$= 6) | • Malaysian aviation industry creates 483 aviation specific job titles in 6 sub-sectors. The highest is the Maintenance, Repairs & Overhaul (MRO) with 279 job titles.<br>• The number of aviation companies increases from 50 (1998) to 266 (2017).<br>• Number of aviation workforce increases from 6,800 (in 1998) to 21,200 (2017) and expected to increase to 32,000 by 2030.<br>• Number of managerial position in Malaysian aviation industry increases from 680 (actual 1998) and 2,120 (actual 2017) to 3,200 (expected 2030), creating shortage of 1,080 of aviation managers. |
| **Stage 2: Exploratory stage** | Thematic analysis | Aviation industry reports & | |

| | | | |
|---|---|---|---|
| *(Industrial sectors level)*<br>To examine the organizational concern with regard to the emergence of Industry 4.0 | | academic literature on IR4.0 in Aviation ($n_{reports}$=8; $n_{articles}$=18) | • Emerging managerial skills related to Industry 4.0 throughout aviation value chain:<br>*Aircraft Design & Certification*: Data mining; Analytical thinking<br>*Aircraft Manufacturing:* Innovative thinking; Innovative analysis; Knowledge in efficiency & effectiveness; Collaborative decision making; Scenario planning; Visualization skill<br>*Airport Operation*: Big Data management; Data driven analytical thinking; Collaborative decision making<br>*Air Traffic Management*: Data driven analytical thinking; Fast data driven & collaborative decision making<br>*MRO*: Real-time data driven decision making; Collaborative decision making<br>*Aviation Training*: Creative thinking; Conceptual thinking |
| **Stage 3: Converging Stage**<br>*(Organizational level)*<br>To examine the employment need on management skills for engineering & technical aviation jobs. | Qualitative interview | Leaders from Malaysian aviation companies ($n_{leaders}$=3) | • Technical aviation professional need to not only technically skilled but also possess managerial skills. E.g. Licensed Aircraft Engineer need to engage into managerial tasks e.g. planning for works & team and motivate them, planning and controlling of budget for manpower & maintenance projects (R1).<br>• With the increase in air traffic that include the UAVs and drones, the Air Traffic Controllers need to acquire analytical thinking and fast decision making skills (R2).<br>• Technical & engineering aviation personnel need to prepare themselves for career switching to management like in the case of Malaysia Airlines Berhad Chief Executive Officer who is a Pilot (R3). |
| **Stage 4: Confirmatory Stage**<br>(Individual level)<br>To verify the individual need for advanced aviation management development | Quantitative Survey | Experienced ($n_{experienced}$=43) & fresh aviation practitioners ($n_{fresh}$=12) | • 93 percent of the respondents agreed obtaining postgraduate aviation management qualification assist both their lateral and vertical career movement.<br>• The urge for enhancing higher academic qualification at postgraduate level is stronger among experienced respondents since they begin to see continuing higher education is also part of organizational need. |

## 4. Conclusion

Providing the opportunity for life-long learning is indeed a major responsibility of any higher education institutions. Yet, offering a customer centric academic program rather than providing 'no- frills' and 'self-indulging' academic program that will not meeting the industry and industry practitioners' expectations and specific needs, is something that a university need to highly ensured. Based on that, this comprehensive market study for the new aviation management postgraduate program was conducted to examine the employability prospects, organizational and personal needs for personal development program, as well as the attractiveness elements of the prospective students. The findings clearly indicate that the prospective graduates of this proposed postgraduate program will have brighter employability changes in aviation industry. This is linked to the high shortage of managerial positions at various management categories that still exists for the next 10 years i.e. until 2030. The emergence of Industry 4.0 technology that encroaches into present aviation industry will also expand the need for certain specific skills related to management such as thinking skills, decision making skills, as well as leading various complex and cross-sectoral teams. The main contribution of this study lies on its methodology. By employing multistages and multilevels, various stakeholders' needs can be appropriately addressed by the higher education institution. This process is important to ensure that each academic program can be self-sustained and continuously relevant in the eyes of the industry's direct and indirect stakeholders.

**Acknowledgement**
The authors are thankful to the Universiti Kuala Lumpur, Malaysian Institute of Aviation Technology (UniKL MIAT) for supporting this research work.

**Declaration of Conflict of Interest**
The author(s) declare no potential conflict of interest with regard to the publication of this research. The list of aviation industry reports and articles (as on Table 2) can be furnished separately upon formal request addressed directly to the corresponding author.

**References**
[1] Department of Skill Development (DSD) & National Aerospace Industry Coordinating Office (NAICO) 2017 *Occupational Framework Aerospace Industry 2017* Department of Skills Development Ministry of Human Resource Putrajaya Malaysia
[2] Liew C L, Talib A A and Jacobs R L 2016 Malaysian aviation technologist promotion to managerial role: An empirical overview *IOP Conf. Ser.: Mater. Sci. Eng.* **152** 012039
[3] Malaysian Aerospace Council (MAC) 2011 *Malaysian Aerospace Industry Report 2011/2012* Malaysian Industry-Government Group for High Technology Selangor Malaysia
[4] Padil H, Said M N and Azizan A 2018 The contributions of human factors on human error in Malaysia aviation maintenance industries *IOP Conf. Ser.: Mater. Sci. Eng.* **370** 1 012035
[5] Magnusson K 1992 Transitions to work: A model for program development. *International Journal for the Advancement of Counselling* **15**(1) pp 27-38
[6] Shaw T, Barnet S, Mcgregor D and Avery J 2015 Using the Knowledge, Process, Practice (KPP) model for driving the design and development of online postgraduate medical education *Medical Teacher* **37**(1) pp 53-58
[7] Wiek A, Withycombe L and Redman C L 2011 Key competencies in sustainability: a reference framework for academic program development *Sustainability Science* **6**(2) pp 203-218
[8] Freeman R E 1984 *Strategic management: A stakeholder approach* Boston Pitman
[9] Leisyte L and Westerheijden D F 2014 *Stakeholders and quality assurance in higher education. In Drivers and barriers to achieving quality in higher education* 83-97 Sense Publishers Rotterdam Netherlands
[10] Fetters M D, Curry L A and Creswell J W 2013 Achieving integration in mixed methods designs—principles and practices. *Health Services Research* **48**(6pt2) pp 2134-2156
[11] Ritter B A, Small E E, Mortimer J W and Doll J L 2018 Designing Management Curriculum for Workplace Readiness: Developing Students' Soft Skills *Journal of Management Education* **42**(1) pp 80-103

[12] Al Ariss A, Cascio WF and Paauwe J 2014 Talent management: Current theories and future research directions *Journal of World Business* **49**(2) pp 173-179

IOP Conf. Series: Materials Science and Engineering **645** (2019) 012009 doi:10.1088/1757-899X/645/1/012009

# Optimization of an Annular Combustion Chamber for Micro Turbo Jet System

**F I Habibi, F Hartono and H Prayogo**

Aero-propulsion Laboratory, Faculty of Mechanical and Aerospace Engineering, Institut Teknologi Bandung, Jalan Ganesha No. 10 Bandung, Indonesia

E-mail: farizichsan4@gmail.com

**Abstract**. This Annular combustion Chamber is designed for micro turbo jet engine. Combustion chamber is positioned in between compressor and turbine of turbojet. Combustion chamber is designed to increase enthalpy without significant pressure loss. Existing combustion chamber is designed by conventional method of trial and error. To further increase its performance then a optimization is needed. The geometry of the combustion chamber is optimized to maximized efficiency, minimized Pattern Factor, minimized pressure loss and lower temperature on the combustion wall. Varying Parameter is chosen based on Sensitivity Analysis with Spearman Correlation method to choose influential changes on geometry. After sampling process with Latin Hypercube Sampling (LHS) method and followed by interpolation with Kriging method, the combustion chamber then optimized using Multi Objective Genetic-Algorithm (MOGA). Performance of the combustion chamber then evaluated by numerical simulation. The combustion chamber will be chosen if it fulfilled pressure loss and temperature on wall criteria. Final design give better performance with a little increase in efficiency and good Pattern Factor.

## 1. Introduction

The combustion chamber is part of a turbo jet that converts chemical energy into enthalpy. A good combustion chamber can burn the right amount of fuel. The correct combustion of fuel will provide enough power so that the motor can operate in accordance with the design. The aero propulsion laboratory has made several turbojet prototypes with various applications. The results of the combustion chamber [1, 2] are used in one combustion chamber test and have a fairly close result between the design and experimental results. When the combustion chamber is added with a compressor and turbine the combustion chamber becomes less stable and easily dies. By using the ECU to facilitate controlling the engine, it increases stability and have better reliability. Another problem arises when the engine operates on a large air discharge the engine easily dies when the throttle is too quickly lowered.

The fire detaches from the swirler and dies if the throttle lowered too fast. To solve this problem, a new annular-shaped combustion chamber is designed. Reverse annular combustion chambers generally have better stability. The air coming in from the front immediately hit the fuel nozzle, so that at high or low speeds the fire would be maintained and not too shifted out. Longer streamlined lines also provide a better pattern factor, see Fig. 1.

IOP Conf. Series: Materials Science and Engineering **645** (2019) 012009 doi:10.1088/1757-899X/645/1/012009

**Figure 1.** Rendered annular reverse combustion chamber

Many considerations in the design of the combustion chamber make the optimum design that difficult to achieve. The design method that has been used was trial and error method [3], so that the design direction is based on the analysis of phenomena in several parts of the combustion chamber. In the real case, the modification in one section will change the entire combustion phenomenon so that limited analysis often not produce an optimum design. To facilitate the design process and reduce design time, writer used MOGA method which is paired with the kriging method. MOGA is used by utilizing genetic algorithms to find the desired optimum point.

## 2. Optimization of Combustion Chamber

The optimization process of combustion chamber performance is based on the flow chart as shown on Fig. 2 below.

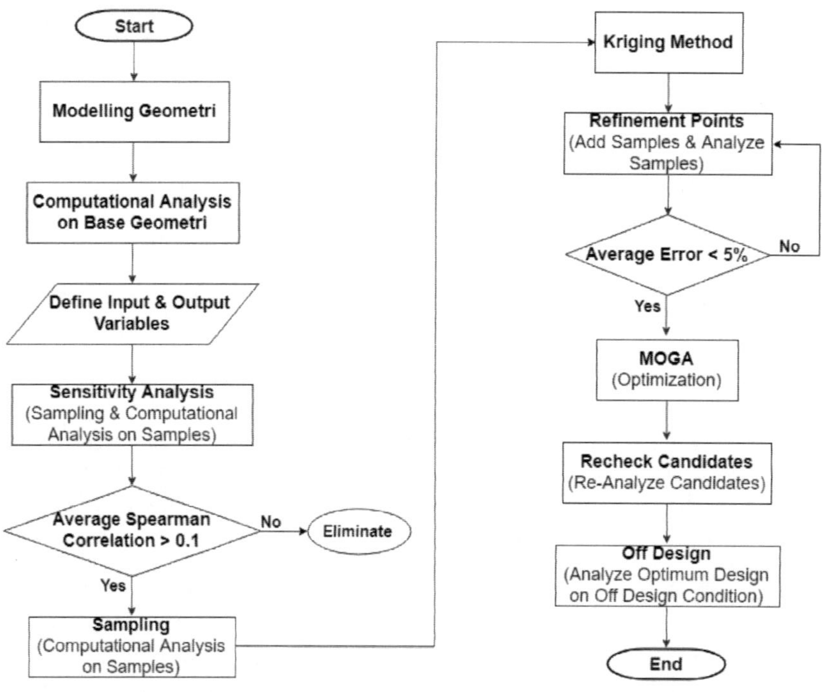


**Figure 2.** Optimization flow chart

Optimization is done using MOGA. Some geometry variables are varied and evaluated for performance using numerical methods. To make the optimization process efficient, sensitivity analysis is carried out. Sensitivity analysis is done to find the relationship between the influence of input variations (geometry variables) on the performance of combustion chamber. For inputs that are not too influential, they will be eliminated from the optimization input. The optimization process that uses genetic algorithms will evaluate many points (individuals) in the design space to get the best optimum point. Evaluation of the performance of the combustion chamber with a numerical method takes quite a long time, therefore, in doing optimization, the kriging method is used to predict the performance of the entire design space. The kriging method will interpolate the sampling results using Latin Hypercube Sampling as many as 300 samples, see Fig. 3.

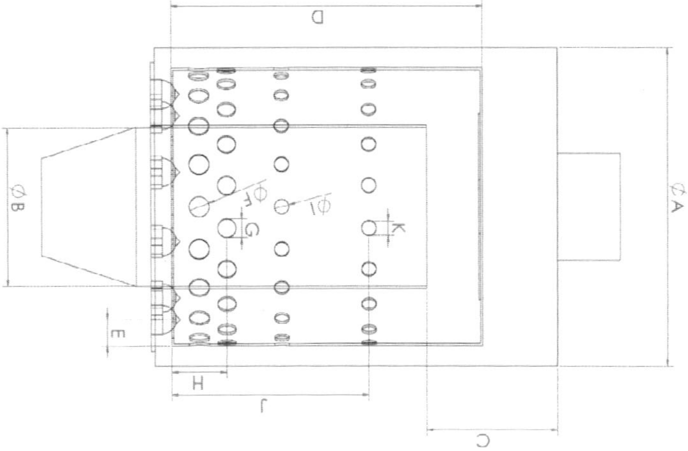


**Figure 3.** Annular reverse combustion chamber notation

*2.1. Sensitivity Analysis*

Sensitivity analysis in this study uses the Spearman correlation method. The Spearman correlation method will measure the monotonic relationship between two variables measured by an ordinal scale. The spearman coefficient(ρ) has a value of -1 to +1. The magnitude of the spearman correlation value indicates the magnitude of the input effect on the output value. The input geometry variable will be eliminated when the average spearman correlation value is less than 0.1. The Spearman correlation value is obtained by the equation below and Table 1.

$$\rho = 1 - \frac{6\sum d_i^2}{n(n^2 - 1)}$$

**Table 1.** Spearman correlation result

| No | Variable | Efficiency | Pattern Factor | Pressure Loss | Tmax Wall | Average Correlation |
|---|---|---|---|---|---|---|
| A | Casing Diameter | 0.300 | 0.466 | -0.594 | 0.564 | 0.481 |
| B | Annulus-Casing Spacing | -0.227 | 0.392 | -0.395 | 0.084 | 0.274 |
| C | 4th Hole Diameter | 0.189 | 0.353 | -0.153 | 0.333 | 0.257 |
| D | Liner Length | -0.238 | 0.325 | -0.322 | 0.123 | 0.252 |
| E | Annulus Diameter | -0.032 | -0.288 | -0.34 | -0.213 | 0.218 |
| F | Atomizer Position | 0.056 | 0.154 | 0.02 | -0.254 | 0.121 |
| G | 2nd Hole Diameter | 0.081 | 0.108 | -0.134 | 0.054 | 0.094 |
| H | 1st Hole Diameter | -0.054 | 0.103 | -0.032 | -0.09 | 0.07 |
| I | 3rd Hole Diameter | 0.149 | 0.040 | 0.022 | 0.051 | 0.065 |
| J | 4th Hole Spacing | 0.053 | 0.026 | -0.032 | -0.021 | 0.033 |
| K | 2nd Hole Spacing | 0.011 | 0.061 | -0.0008 | -0.047 | 0.03 |

### *2.2. MOGA*

Multiple Objective Genetic Algorithm (MOGA) is a type of NSGA II method (non-dominated sorted genetic algorithm II) to find the optimum point based on the desired target where more than one desired objective. Because the desired target is 2 objectives, a fitness function is needed to provide the desired objective-weighting, as presented on Table 2.

**Table 2.** MOGA Target and constrain

| | | |
|---|---|---|
| Efficiency | Objective | High priority |
| Pattern factor | Objective | High priority |
| Pressure loss ratio | Constraint | less than 0.2 |
| Tmax Wall | Constraint | less than 1300 K |

Workflow from MOGA begins with random generation of initial populations taken from the response surface that generated by kriging interpolation. Furthermore, from the initial population it will produce new generations generated through the process of genetic algorithms in which there is a selection process, cross-section and mutation to produce new individuals. The formation process of this generation will continue until it reach the specified convergence critical limits. Convergence criteria are used in this optimization is the stability percentage convergence and the maximum number of generations. The stability percentage convergence (algorithm convergence) is a percentage which represents the stability of the population, based on its mean and standard deviation [4, 5].

Maximum number of generations is the maximum possible number of generations the algorithm executes. If this number is reached without the optimization having reached convergence, iteration will stop. This also provides an idea of the maximum possible number of function evaluations that are needed for the full cycle, as well as the maximum possible time it may take to run the optimization, as presented in Fig. 4.

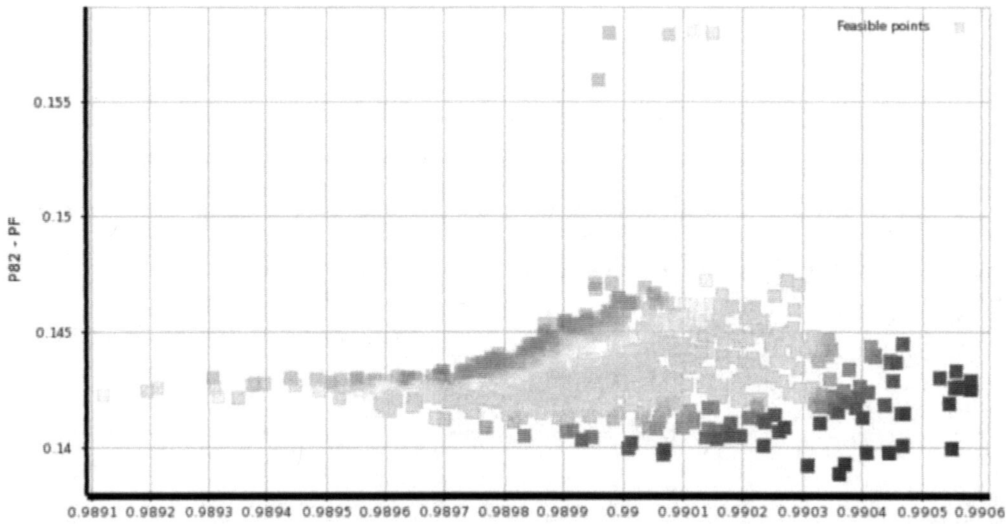


**Figure 4.** Pareto Frontier

IOP Conf. Series: Materials Science and Engineering **645** (2019) 012009 doi:10.1088/1757-899X/645/1/012009

## 3. Results and Discussions

Performance form kriging interpolation produced satisfying result. This could achieve higher efficiency, and lower pattern factor which are desired. However, combustion chamber produced higher pressure loss and higher temperature on the wall. Since it just an idea we have to check it again with CFD analysis not with interpolation but real geometry, highlighted in Table 3 and Fig. 5.

**Table 3.** Optimized combustion chamber performance

| Performance | Base | Optimized | Delta (%) |
|---|---|---|---|
| Efficiency | 93.742 | 98.675 | 4.933 |
| Pattern factor | 0.927 | 0.59 | -36.356 |
| Pressure loss | 0.176 | 0.191 | 8.991 |
| Tmax Wall | 1161.6 | 1255.15 | 8.054 |

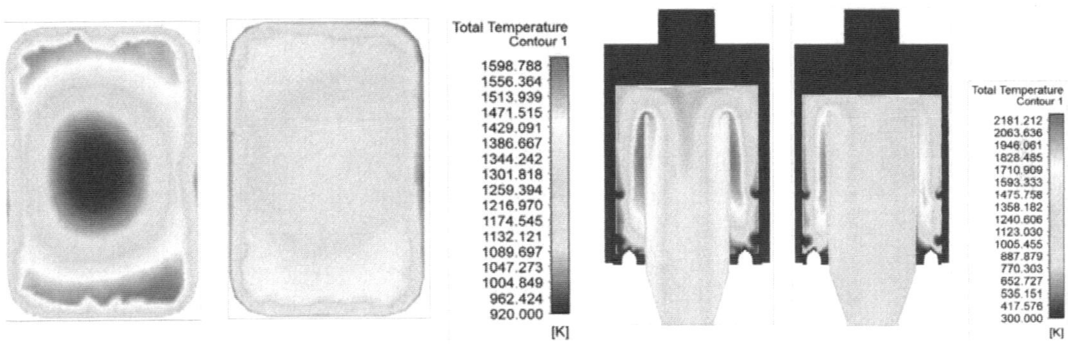


**Figure 5.** Total temperature contour

The temperature contour from optimization result is more evenly distributed after passing through the liner area and entering the outlet channel, compared to base geometry. In the outlet section, it also have more distributed temperature and the value of the pattern factor is smaller than the base geometry result. From the temperature contour, it can also be seen that the temperature on the combustion chamber wall have smaller area with maximum temperature. Temperature on the outlet wall also lower than the base geometry. This occurs due to changes in some geometry size, including the reduction of the width of the annulus and the position of the atomizer away from the inner liner wall. Where, the inner liner wall is the place of the maximum temperature in combustion chamber.

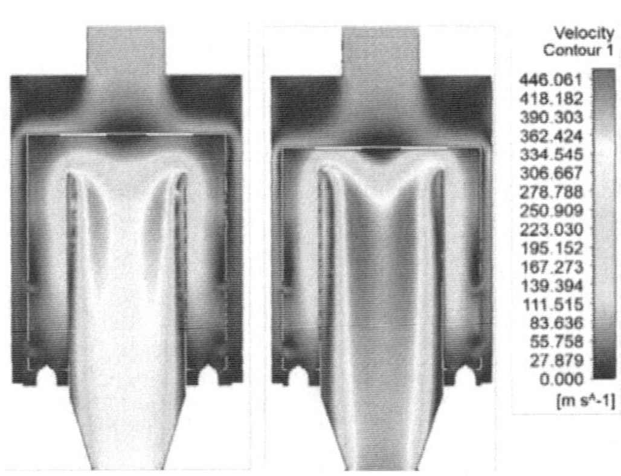


**Figure 6.** Velocity contour

IOP Conf. Series: Materials Science and Engineering **645** (2019) 012009 doi:10.1088/1757-899X/645/1/012009

There is some reduction in the area of the annulus that causes the velocity in the annulus area increase. The increase in annulus velocity is followed by an increase in the penetration angle of the liner holes [5, 6]. With increasing penetration angle of the liner holes increase, the intensity of turbulence inside the liner will increase and produce a better mixing quality [6]. The efficiency of combustion will increase by increasing the quality of the mixture so that the optimization results of the reverse annular combustion chamber have an efficiency increase by 2.3%.

Decreasing on pattern factors and increasing on efficiency after being optimized is followed by an increase in pressure loss in this combustion chamber. Pressure loss is generated due to the skin friction on the combustion chamber wall and the separation flow in the combustion chamber. The pressure loss of the annular reverse combustion chamber is quite large, because of the flow pattern that turns $180^0$ twice, so it will produce the separation flow and a larger skin friction. It increase pressure loss mainly because lesser annulus area, with casing and liner wall getting closer it will increase shear stress. Interaction between wall become more apparent and made higher pressure losses, see Table 4.

**Table 4.** Combustion Chamber performance comparison

| Parameters | Base Geometry | Optimized Geometry | Delta (%) |
|---|---|---|---|
| Efficiency (%) | 93.58 | 95.90 | 2.315 |
| Pattern Factor | 0.412 | 0.181 | -56.07 |
| Pressure loss ratio | 0.151 | 0.190 | 20.52 |
| Tmax on the wall (K) | 1302.07 | 1242.03 | -4.611 |

## 4. Conclusions

After analyzing optimized geometry of the combustion chamber, it has better performance. As already mentioned before, there're slight decrease in performance aspect but not overshadowing the increased efficiency and pattern factor. In term of percentage we have 20% higher pressure loss, but in number it is around 0.039 which is still acceptable. The higher efficiency, lower pattern factor, lower temperature on combustion chamber wall and generally increases in performance. It could be achieved a better performance with MOGA to design combustion chamber. For future work writer suggest to optimized engine as a whole turbo jet. Which would increase performance not just each part but overall performance as a whole engine. Maybe there could be some interaction between part that still uncertain, and by doing optimization to whole system it may have better understanding of those interaction.

## References

[1] Fuad N, Hartono F and Moelyadi M A 2015 *Combustor Design for Turbocharger Turbojet Engine*, International Conference on Intelligent Unmanned System (Bali)
[2] Hartono F 2015 *The Development of Turbojet Based on Turbocharger*, International Conference on Intelligent Unmanned System (Bali)
[3] Lefebvre, A. H. 1983. Gas Turbine Combustion. Hemisphere Publishing Corp. New York.
[4] Cohen, Henry. *Gas Turbine Theory*. Longman Group Limited, England.
[5] Farokhi, Saeed, 2014. *Aircraft Propulsion and Gas Turbine Engines*, John Wiley & Sons, United Kingdom
[6] Kaddah, K.S. Discharge Coefficient and Jet Deflection Angles for Combustor Liner Air Entry Holes, College of Aeronautics M.Sc thesis, Cranfield, England, 1964.

IOP Conf. Series: Materials Science and Engineering **645** (2019) 012010 doi:10.1088/1757-899X/645/1/012010

# Analytical assessment on aviation practitioners needs for knowledge enhancement at the postgraduate level: an application of Fuzzy Delphi method

**M F N Mohammad[1,2,*], N H Ghafar[1], N A Rahman[1], R Abdullah[2] and F Mayor[3]**

[1]Malaysian Institute of Aviation Technology, Universiti Kuala Lumpur, Selangor, Malaysia

[2]Faculty of Computer Sciences and Information Technology, Universiti Putra Malaysia, Selangor, Malaysia

[3]Department of Economics and Social Sciences, Universitat Politecnica De Valencia, Valencia, Spain

*E-mail: fakhrulnizam@outlook.com

**Abstract.** Knowledge enhancement is essential to support continuous sustainability of organization in a competitive environment. It is not an exception to an aviation industry. In facing up with future challenges of industrial revolution 4.0 (IR 4.0), studies have shown that one of the issues addressed in ensuring it success is the importance of knowledge among employees. Various activities can be done to enhance knowledge such as by enrolling to academic (undergraduate or postgraduate) programs, attending short courses/training or sitting for professional certification. Furthermore, obtaining postgraduate degrees offers wide exposure to academic and industry environment. Therefore, the purpose of this study is to gauge the understanding on practitioners' key reasons to further their education at the postgraduate level. Four major constructs established for this study to analyse the aviation practitioners' intention to further study that are personal needs for continuing education, organization needs, source of funding and teaching and learning facilities. Document review were conducted at the beginning of the study to explore fundamental issues of knowledge enhancement needs among practitioners, meanwhile survey performed were analysed by using Fuzzy Delphi method (FDM). Findings shows that six elements (items) for the whole survey fulfil the criteria of threshold value (*d* construct) less than 0.2 and achieved the percentage of expert group consensus more than 75%. This research provides objective and constructive input to university on the major factors influencing aviation practitioner's decision to further their study at the postgraduate level. The finding becomes avenues for university to draft a better postgraduate program to fulfil the requirement of industry and academic.

## 1. Introduction

Knowledge and skilful workers are the catalyst for the development of a country [1]. Pursuing formal education at postgraduate level is a continuous improvement that helps many to improve and enhance fundamental and applied knowledge. However, there are other factors that may affect the choices of pursuing study among practitioners. External and internal factors play important role in influencing the practitioners' decision to pursue their study at postgraduate level. Within education sector that specialises on aviation program, there are very minimal number of postgraduate programs offered by university.

It is even more difficult to find aviation management program courses offered not only within Asia but in another region as well. Scarcity of programs may become external factors that deter aviation practitioners to further studies at postgraduate level. On the other hand, internal factors that gearing them up to further their studies are also important factors to look at. Furthermore, studies on practitioners' decision to pursue their study at the postgraduate level remains unexplored. Therefore, studies need to be conducted in identifying the factors that influences practitioners' decision in enhancing knowledge and pursue their study at postgraduate level.

Development of aviation management program at postgraduate level is hoped to fill up the gaps in offering aviation practitioners' opportunities to upgrade their academic qualification. Development of master's degree in aviation management is to support the development of future managers in the industry. It is estimated that by year 2030, Malaysia requires a total number of 32,000 aviation workforce [2]. From this figure, there is a conjecture for the compounding needs of numbers of managers and senior executive that requires higher academic qualification.

Malaysia has about 53 education and training institute in aerospace related programmes comprises of 27 higher education institutions, 15 technical academies and 11 DCA (Part 145) approved training organizations [3]. However, the dearth of management program in aviation offered is noteworthy; there is only one bachelor's degree in aviation management offered currently, and there is no masters' degree in aviation program offered yet. There is other aviation related program at postgraduate level currently offered in the other region such as Embry Riddle University (United States of America), Cranfield University (United Kingdom) or in Emirates Aviation University (Dubai) but not specifically to aviation management program. In year 2014, 3,637 skilled workers have been produced within South East Asia; Malaysia alone has produced about 54% of the skilled workers within region [3]. In the Asia Pacific region, Malaysia is forecasted to be the forefront in aviation sector. Furthermore, education and training sector are considered as essential in supporting the initiatives as highlighted both in the Malaysia Aerospace Industry Blueprint 2015-2030.

In realizing the emergence needs of an aviation management program at postgraduate level, an online survey was conducted to practitioners in aviation industry; to gauge their feedback for continuing education at the postgraduate level. The survey questionnaire has been distributed among practitioners' in Malaysia. The major constructs of the survey questions using multi-stage multi-level analysis that comprises of four categories: personal needs for education, organizational needs, sources of funding and teaching and learning facilities. From the education provider perspective, identifying the main objective for prospective students (practitioners) to pursue their academic qualification at the postgraduate level is a golden quest. Since the purpose of this study is to conduct an analytical analysis on respondents' survey, therefore this study will provide a guide for other researcher in conducting research survey based on FDM. This study contributes for more discussion on FDM use in analysing the survey respondents.

## 2. Fuzzy Delphi method

Multicriteria decision making (MCDM) tools such as Analytical Hierarchy Process (AHP), Delphi method, Technique for Order of Preference by Similarity to Ideal Solution (TOPSIS), VIKOR method and others have been widely used in organization as well as in academic studies. The purpose of using MCDM tools are normally because these tools have been incorporated with consistency checking [4] therefore minimizes subjective influences in decision making [5].

Delphi method developed in 1950s by RAND corporation [6], provide benefits in assessing decision such as: (1) effectiveness of decision making in the absence of past data. (2) survey can be made remotely without face to face and (3) its usefulness in future prediction for new technology or product [7]. However, there are also critiques on the weaknesses of Delphi method as discussed by [7]. Combination of fuzzy set theory and Delphi method introduced by [8] provide stability in the result obtained from analysis.

Study by [9] defined different roles of MCDM tools such as FDM, FAHP and Fuzzy TOPSIS and its application. For instance, FDM is used to determine the important criteria for building the hierarchy in MCDM. Subsequently, FAHP is adopted to elaborate the weight of criteria and fuzzy TOPSIS on the other hand deployed to rate alternatives used in decision making made by decision maker. The practices adopted by [9] accentuating earlier practices discussed by [10]; that adopts the use of modified Delphi in determining and analysing criteria based on the conducted survey. The relation between these three decisions making tools are illustrated by [9] in Figure 1.

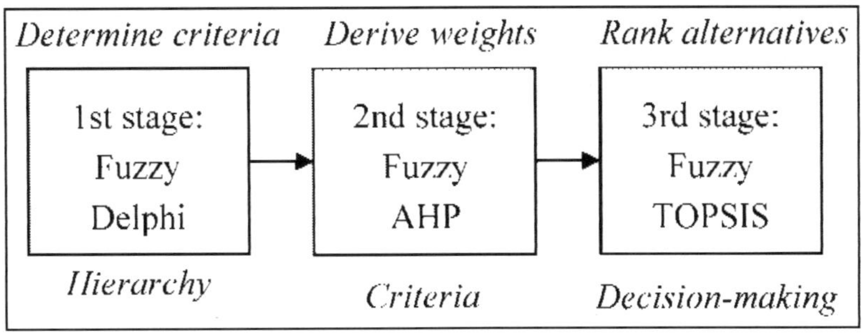


**Figure 1**: The three-stage technology selection process

Figure 1 indicated that FDM were assigned to determine criteria and develop the hierarchy prior to further process by FAHP and Fuzzy TOPSIS. The same practice are used earlier by [10] that adopts modified Delphi prior to analysis by FAHP. Similarly, FDM are used to explore important criteria in selection [7,11] and to develop primary evaluation criteria [12,13]. Therefore, we suggest the use of FDM to analyse questionnaire survey conducted among aviation practitioner. Furthermore, most suitable collection of information via group decision making and feedback from experts are to use Delphi method [14]. Survey result were analysed using FDM by performing analytical assessment to investigate the needs of knowledge enhancement among aviation practitioners will witness a concrete and stability in term of the findings and reporting.

## 3. Methodology

Purpose of this study is to analyse the result obtained from survey conducted online. Survey is chosen as the strategy of inquiry to get feedback from aviation practitioners with regards to the needs of enhancing knowledge; by pursuing in the academic program at the postgraduate level (master's degree). Survey questionnaire were developed on four major constructs built for this study represented in the following Figure 2. Seven items Likert scale are used to acquire respond from the respondent, however the preferential levels are differently used for each question as shown in Table 1.

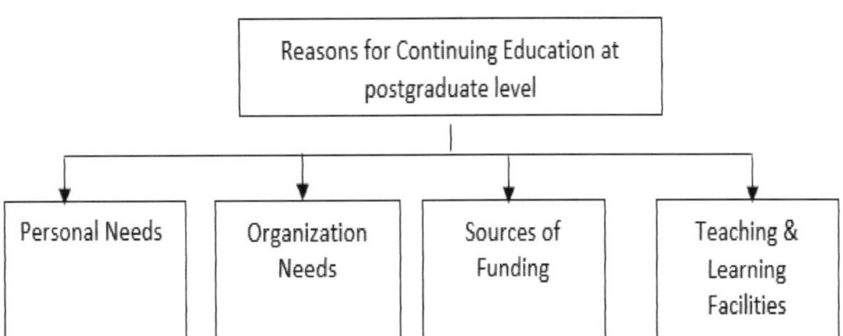


**Figure 2**: Major constructs built for this study

IOP Conf. Series: Materials Science and Engineering **645** (2019) 012010 doi:10.1088/1757-899X/645/1/012010

Survey was conducted online and distributed through referral and contacts within a month using an online form. The advantages of using online survey is costs saving, shorter time to acquire data, access to unique and unreachable population [15] and easier to reach remote users [16]. At the end of the survey period, 55 respondents have responded and answered all questions. The number of respondents is considered enough for the time being; considering the period for the survey (within a month) and has expedited efforts to promote the survey (through referral). Nonetheless, there are still arguments and discussions over the minimum number (15 respondents) and maximum number (50 respondents) for Delphi method [17].

The list of preferential level is shown in Table 1. In overall, there are seven types of preferential level used in this survey. Subsequently, seven items Likert scale and seven points FDM as shown in Table 2 were matched and converted. The whole FDM process was performed using Microsoft Excel 2016. Template for calculating FDM result were developed. The survey result was analysed and included in Microsoft Excel. At the end of the process, findings from the analysis were conducted and reported. The findings of the analysis are discussed in the next section. Analysis of the result for this study was conducted in two stages as described in Figure 3.

**Table 1:** Preferential Table Used in the Survey Questionnaire

| Suitability Level | Importance Level | High Level | Likeliness Level | Preferential Level | Requirement Level | Agreement Level |
|---|---|---|---|---|---|---|
| Strongly Suitable | Extremely Important | Extremely High | Strongly Likely | Highly Preferred | Highly Required | Strongly Agree |
| Mostly Suitable | Mostly Important | Most Likely High | Most Likely | Most Preferred | Slightly Required | Mostly Agree |
| Suitable | Important | High | Likely | Preferred | Required | Agree |
| Neutral | Neutral | Neutral | Neutral | Neutral | Neutral | Neutral |
| Not Suitable | Not Important | Lower | Unlikely | Non-Preferred | Not Required | Disagree |
| Mostly Not Suitable | Mostly Not Important | Mostly Lower | Most Unlikely | Most Non-Preferred | Mostly Not Required | Mostly Agree |
| Strongly Not Suitable | Strongly Not Important | Extremely Lower | Strongly Unlikely | Strongly Non-Preferred | Strongly Not Required | Strongly disagree |

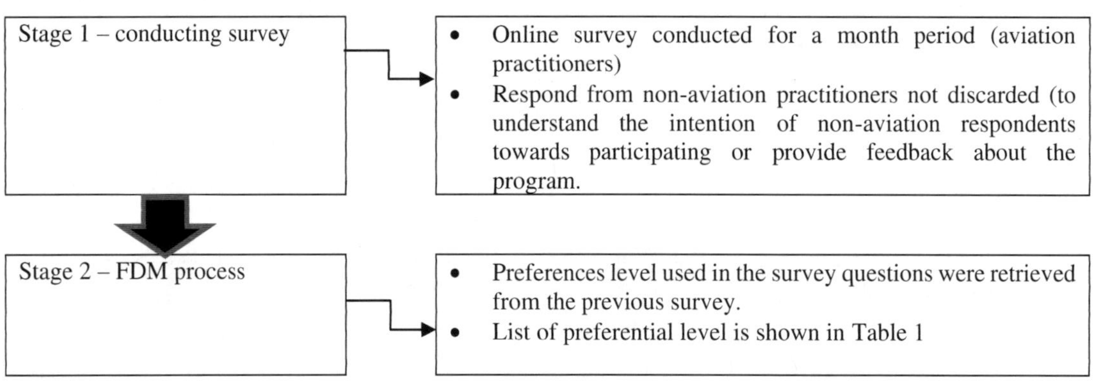


**Figure 3** : Analysis steps

**Table 2 :** Matching Table between Likert scale and 7 Point Fuzzy scale
Adopted from [18,19]

| Preferential Level | 7 points Fuzzy Scale | Likert Scale |
|---|---|---|
| Strongly disagree | (0.0,0.0,0.1) | 1 |
| Somewhat disagree | (0.0,0.1,0.3) | 2 |
| Disagree | (0.1,0.3,0.5) | 3 |
| Neutral | (0.3,0.5,0.7) | 4 |
| Agree | (0.5,0.7,0.9) | 5 |
| Somewhat agree | (0.7,0.9,1.0) | 6 |
| Strongly agree | (0.9,1.0,1.0) | 7 |

## 4. Analysis and Findings

Results of the survey shows that majority of the respondents from this study are currently an active employee working within aviation industry in Malaysia. Meanwhile, employees from other industry and current students in aviation/aerospace sectors were also involved and provide their feedback in the study. In term of the jobs\employment level as shown in Table 3, majority of respondents are from the executive\senior executive\engineer\officer categories. Second in place, the total number of 15 respondents such as students\lecturer\SL1M (*Skim Latihan 1 Malaysia*\ Malaysia Training Scheme program) trainees also participated in the study apart from employee at the managerial\directors' level.

**Table 3** : Types of Respondents' Jobs

| No | Jobs level\areas | Example | Quantity |
|---|---|---|---|
| 1 | Manager\Director | Director, Quality Assurance Manager, Human Resource manager, Assistant manager, Freight manager, Ramp manager. | 9 respondents |
| 2 | Executive\Senior Executive\Engineer\Officer | NOTAM officer, Process engineer, Route revenue analyst, Operation executive, Human Resource officer. | 28 respondents |
| 3 | Education within aviation | Lecturer, Students, SL1M trainee | 15 respondents |
| 4 | Others | Pilot, Airforce Inspector, Banking officer | 3 respondents |
| | **Total respondents** | | **55 respondents** |

From the business segments of the respondents, Figure 4 shows that majority of the respondents comes from the operations\engineering\technical area within several domains in the aviation sector such as airlines operations, airport, air-traffic controller, maintenance, repair & overhaul (MRO). This figure will lead university for detailed study in future to identify the exact target prospective students for more relevant syllabus preparation.

FDM was conducted on the result from 27 items of survey response; based on the survey questionnaire obtained earlier. From the 29 survey questions, two questions were discarded as both questions are descriptive in nature.

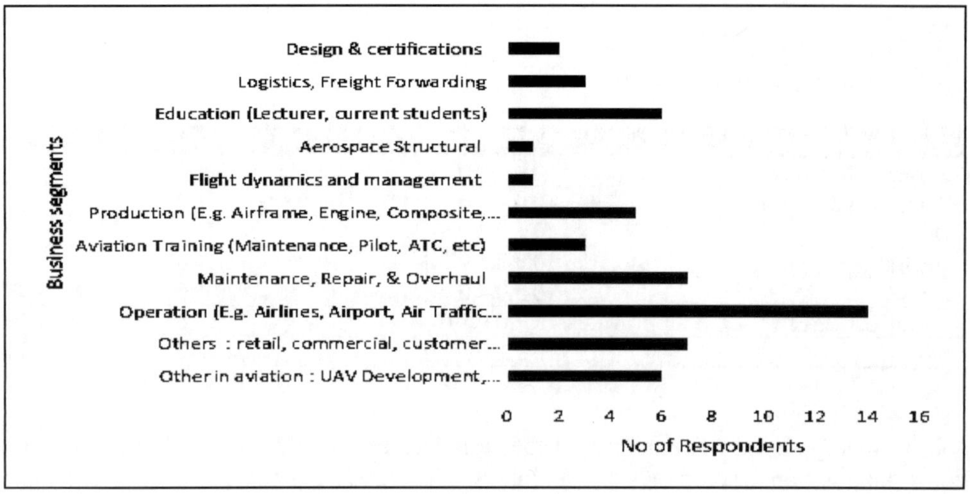


**Figure 4:** Categories of Business Segments

*4.1 Conditions and Percentage*

FDM outlines two types of conditions needs to be qualified and fulfilled by the result: threshold value *d* and percentages of expert's consensus of more than 75% [1]. Survey questionnaire was clustered from four major constructs defined earlier that are personal needs for continuing education, organizational needs, source of funding and teaching & learning facilities. From Table 4, FDM processes shows that six items (elements) are accepted after getting threshold value less than 0.2 and accepted in consensus by the experts after getting the percentage of consensus more than 75%.

**Table 4 :** FDM process result

| No | Elements | Triangular Fuzzy Numbers | | Defuzzification Process | | | | Expert Consensus | Ranking |
|---|---|---|---|---|---|---|---|---|---|
| | | Threshold Value, d | Expert Consensus, % | m1 | m2 | m3 | Fuzzy Score (A) | | |
| 1 | A1 | 0.3 | 80% | 0.658 | 0.811 | 0.902 | 0.790 | ACCEPT | 11 |
| 2 | A2 | 0.2 | 93% | 0.718 | 0.873 | 0.953 | 0.8479 | ACCEPT | 4 |
| 3 | A3 | 0.2 | 96% | 0.747 | 0.896 | 0.969 | 0.8709 | ACCEPT | 3 |
| 4 | A4 | 0.3 | 78% | 0.638 | 0.802 | 0.904 | 0.7812 | ACCEPT | 13 |
| 5 | A5 | 0.3 | 84% | 0.736 | 0.833 | 0.918 | 0.8291 | ACCEPT | 6 |
| 6 | A6 | 0.3 | 80% | 0.651 | 0.813 | 0.911 | 0.7915 | ACCEPT | 10 |
| 7 | A7 | 0.2 | 91% | 0.835 | 0.724 | 0.851 | 0.9018 | ACCEPT | 1 |
| 8 | B8 | 0.3 | 47% | 0.555 | 0.802 | 0.902 | 0.7097 | REJECT | 16 |
| 9 | B9 | 0.3 | 80% | 0.640 | 0.769 | 0.887 | 0.7812 | ACCEPT | 13 |
| 10 | B10 | 0.3 | 47% | 0.598 | 0.809 | 0.911 | 0.7515 | REJECT | 15 |
| 11 | B11 | 0.3 | 82% | 0.647 | 0.809 | 0.911 | 0.7891 | ACCEPT | 12 |
| 12 | C12(a) | 0.3 | 42% | 0.216 | 0.365 | 0.540 | 0.3739 | REJECT | 26 |
| 13 | C13(a) | 0.5 | 13% | 0.382 | 0.518 | 0.651 | 0.5170 | REJECT | 24 |
| 14 | C14(a) | 0.4 | 27% | 0.531 | 0.687 | 0.805 | 0.6745 | REJECT | 19 |
| 15 | C15(a) | 0.4 | 33% | 0.373 | 0.533 | 0.685 | 0.5303 | REJECT | 22 |
| 16 | C16(desc) | | | Non rating questions | | | | | |
| 17 | D17(a) | 0.4 | 24% | 0.275 | 0.442 | 0.625 | 0.4473 | REJECT | 25 |
| 18 | D18(a) | 0.4 | 38% | 0.342 | 0.522 | 0.704 | 0.522 | REJECT | 23 |
| 19 | D19(a) | 0.3 | 35% | 0.436 | 0.627 | 0.793 | 0.6188 | REJECT | 20 |
| 20 | D20(a) | 0.2 | 87% | 0.518 | 0.715 | 0.878 | 0.7036 | ACCEPT | 17 |
| 21 | D21(b) | 0.3 | 85% | 0.669 | 0.813 | 0.902 | 0.7945 | ACCEPT | 9 |
| 22 | D22(b) | 0.3 | 78% | 0.678 | 0.827 | 0.918 | 0.8079 | ACCEPT | 8 |
| 23 | D23(b) | 0.4 | 85% | 0.716 | 0.845 | 0.913 | 0.8248 | ACCEPT | 7 |
| 24 | D24(b) | 0.2 | 91% | 0.775 | 0.902 | 0.955 | 0.8770 | ACCEPT | 2 |
| 25 | D25(b) | 0.2 | 93% | 0.733 | 0.869 | 0.938 | 0.8467 | ACCEPT | 5 |
| 26 | D26(c.) | 0.4 | 47% | 0.433 | 0.602 | 0.756 | 0.5970 | REJECT | 21 |
| 27 | D27(c.) | 0.3 | 65% | 0.522 | 0.687 | 0.827 | 0.6788 | REJECT | 18 |
| 28 | D28(c.) | 0.3 | 80% | 0.642 | 0.795 | 0.895 | 0.7770 | ACCEPT | 14 |
| 29 | D29(desc) | | | Non rating questions | | | | | |
| | TOTAL ITEMS | 6 | 16 | | | | | 6 | |

Threshold value *d* and Consensus Percentage are conditions used in FDM to display the consensus among experts on the items agreeable in the study; of which the *d* value must be less than 0.2 and the group of experts consensus must be more than 75% [20]. Inability to get the required value and percentage, items either needs to be removed or require second round of FDM. Defuzzification process were constructed to identify the rank of elements (importance) of respondents' selection over the decision on enhancing their knowledge at the postgraduate level (master's degree).

*4.2 Threshold Value and Consensus Percentage*

This study utilizes questionnaire survey that comprises of 29 questions. However, only 27 questions were analysed using FDM; the other two questions were descriptive in nature. Table 5 shows only six elements are accepted which those questions are getting *d* value less than 0.2 are items under the personal needs in continuing education and teaching and learning facilities.

Aviation practitioners decided to pursue their study due to their personal needs in enhancing knowledge. This has some reflection to the descriptive figure on the type of respondents' level of jobs that majority of them are coming from executive\senior\executive\officer level; as opposed to other types of categories. It briefly concluded that the quest for knowledge enhancement among this group is relatively high. It also enlightened that most respondents believe that postgraduate program (masters' degree) will enhance their managerial skills, they will feel satisfied if they are able to upgrade academic qualification and belief that age does not an issue for continuing education.

On top of their personal factors that influences the decision, second factors related to teaching and learning facilities provided by the university are also important. The result shows that, prospective students appreciated for flexible learning and more practical types of learning as indicated by their choices of prefer classes on modular basis, availability of simulated learning facility and preference of access to e-Resources (learning materials accessible through online\internet). Most 91% of respondents prefer classes on modular basis and having access to e-resources. One of the major significant finding is that, the respondent also prefers the availability of simulated learning facility to combine between academic build up and the industry experience. Therefore, they would be able to link immediately with the experience in the real-life situation.

**Table 5** : Result of Fuzzy Delphi Method on respondent priority of continuing education

| No | Code | Accepted Item | Threshold d value | Expert Consensus Percentage |
|---|---|---|---|---|
| 1 | A2 | Master program will enhance managerial skills | 0.2 | 93% |
| 2 | A3 | Feel satisfied if able to upgrade academic qualification | 0.2 | 96% |
| 3 | A7 | Age does not stop from upgrading education. | 0.2 | 91% |
| 4 | D20(a) | Prefer classes on modular basis | 0.2 | 87% |
| 5 | D(24(a) | Prefer access to e-Resources | 0.2 | 91% |
| 6 | D25(a) | Existence of simulated learning facility | 0.2 | 93% |

*4.3 Ranking of Items*

Table 6 and Table 7 shows the ranking of accepted and rejected items based on the consensus from the expert responds. It is to note that most experts in consensus agreed that personal needs and teaching and learning facilities influences main reason for practitioners in pursuing education. Source of funding for prospective students may not be the main reason at the time the questions was answered. Main reasons

IOP Conf. Series: Materials Science and Engineering **645** (2019) 012010 doi:10.1088/1757-899X/645/1/012010

could be: 1) fees amount was not mentioned in the study and 2) respondents were asked in term of their source of funds but not their capabilities on how to pay for the fees.

**Table 6** : Ranking of items – Expert Consensus > 75% (Accepted)

| Ranking of items | Fuzzy Score | Accepted Item | Expert Consensus | Category |
|---|---|---|---|---|
| 1 | 0.902 | Age does not stop from upgrading education. | 91% | PN |
| 2 | 0.877 | Prefer access to e-Resources | 91% | TL |
| 3 | 0.871 | Feel satisfied if able to upgrade academic qualification | 96% | PN |
| 4 | 0.848 | Master program will enhance managerial skills | 93% | PN |
| 5 | 0.847 | Existence of simulated learning facility | 93% | TL |
| 6 | 0.829 | Masters' degree enables to move to another domain | 84% | PN |
| 7 | 0.825 | Access to campus WiFi | 85% | TL |
| 8 | 0.808 | Must have separate training for study skills | 78% | TL |
| 9 | 0.795 | Access to virtual leaning platform | 85% | TL |
| 10 | 0.792 | Master's degree able to increase promotion | 80% | PN |
| 11 | 0.790 | Aim to enrol in master's degree in aviation management related | 80% | PN |
| 12 | 0.789 | People will higher education level may have career mobility | 82% | ON |
| 13 | 0.781 | Master's degree able to move upward in organization hierarchy | 78% | PN |
| 13 | 0.781 | Organization encourage staff to involve in self-development program | 80% | ON |
| 14 | 0.777 | Blended between academic research and industry case study | 80% | TL |
| 17 | 0.704 | Prefer classes on modular basis | 87% | TL |

**Table 7:** Ranking of Items - Expert Consensus < 75% (Rejected)

| Ranking | Fuzzy Score | Accepted Item | Experts Consensus | Category |
|---|---|---|---|---|
| 15 | 0.752 | The higher position for a person, the more it demands for academic qualification upgrade | 47% | ON |
| 16 | 0.710 | Present job demands for more management skills related. | 47% | ON |
| 18 | 0.679 | Prefer for final year project to be purely industry case study. | 65% | TL |
| 19 | 0.675 | Sources of fund from external sponsorship | 27% | SF |
| 20 | 0.619 | Prefer have classes on weekends. | 35% | TL |
| 21 | 0.597 | Prefer for final year project to be purely academic research thesis | 47% | TL |
| 22 | 0.530 | Sources of fund from EPF Withdrawal scheme. | 33% | SF |
| 23 | 0.522 | Prefer classes at night | 38% | TL |
| 24 | 0.517 | Sources of fund from company's sponsorship | 13% | SF |
| 25 | 0.447 | Prefer have classes between 8am to 5pm | 24% | TL |
| 26 | 0.374 | Sources of fund from own saving | 42% | SF |

Categories: PN = Personal Needs, ON = Organization Needs, TL = Teaching & Learning facilities and SF = Source of Funding

## 5. Conclusion

Knowledge can be defined into tacit and explicit knowledge [21]. Acquiring and expanding knowledge is a continuous process whereby enhancement of knowledge can be achieved from formal or informal education. Challenges of universities in the new era is to provide education that can fulfil the needs of industry and produce more knowledge workers. More challenges are foreseeing especially in the upcoming of Industrial Revolution 4.0 (IR 4.0). From the industry standpoint, expansion of intellectual capital assures continuous operations and sustainability of the organization. The analysis used FDM process has successfully defined the main reasons practitioners' pursuing their studies mainly due to their personal needs and teaching and learning facilities offered by university. Constructively, FDM is one of the MCDM tools used to analyse decision and criteria of respondents by delineating biasness, subjective responds and at the same time retrieve consensus agreement among experts. The result of FDM process provides an early understanding for education and training institute on prospective students' intention in continuing their education at the postgraduate level. University as the education provider may prioritise to emphasize on the two categories of concern by prospective students that are their personal needs in continuing education at the postgraduate level as well as to provide conducive teaching and learning facilities. In future, further studies may emphasize on the modification to overcome some of the weaknesses with the questionnaire items and the design. Furthermore, the survey was conducted earlier before engaging into FDM process. FDM questionnaire design might be considered in future while designing the questions for more structured type of questions and analysis.

## References

[1] Hasan A, Hafiz F, and Shahril M 2017 Application of Fuzzy Delphi Approach Determining Element in Technical Skills among Students towards the Electrical Engineering Industry Needs *Pertanika J.Soc.Sci Humanit.*, vol. 25, no. S, pp. 1–8

[2] Mohammad M F, and Rahman N A A 2015 Analysis of Competition Issues for Universiti Kuala Lumpur, Malaysian Institute of Aviation Technology (UNIKL MIAT): Towards Achieving Malaysian Aerospace Industry Blueprint 2015-2030 *Procedia - Soc. Behav. Sci.,* vol. 211, pp. 226–231

[3] Zainal K, Talib A R A, and Hack I 2015 Malaysian Aerospace Industry Blueprint

[4] Mohammad M F, Rahman N A A, Hassan R, and Kurniawan R 2017 Multi-Criteria Decision Making for Bachelor of Aircraft Engineering Technology (Avionic) Final Year Students in project Management Course : AHP Method for Career Selection *J. Eng. Appl. Sci.,* vol. 12, no. 3, pp. 705–714

[5] Mammadova M, and Jabrayilova Z 2014 Application of Fuzzy Optimization Method in Decision-Making for Personnel Selection *Intell. Control Autom.,* vol. 05, no. 04, pp. 190–204

[6] Dalkey N 1969 The Delphi Method: An Experimental Study of Group Opinion *The RAND Corporation.* pp. 408–426

[7] Chang P L, Hsu C W, and Chang P C 2011 Fuzzy Delphi method for evaluating hydrogen production technologies *Int. J. Hydrogen Energy*, vol. 36, no. 21, pp. 14172–14179

[8] Murray T J, Leo L P, and Gigvh J P 1985 A pilot study of fuzzy set modification of Delphi *Hum. Syst. Manag.,* vol. 5, no. 1, pp. 76–80

[9] Lee Y C and Chou C J 2016 Technology evaluation and selection of 3DIC integration using a three-stage fuzzy MCDM *Sustain.*, vol. 8, no. 2, pp. 1–15

[10] Hsu P F and Chen B Y 2007 Developing and implementing a selection model for bedding chain retail store franchisee using Delphi and Fuzzy AHP *Qual. Quant.*, vol. 41, no. 2, pp. 275–290

[11] Hsu Y L, Lee C H, and Kreng V B 2010 The application of Fuzzy Delphi Method and Fuzzy AHP in lubricant regenerative technology selection *Expert Syst. Appl.*, vol. 37, no. 1, pp. 419–425

[12] Cheng J H, Lee C M, and Tang C H 2009 An application of fuzzy delphi and fuzzy AHP on evaluating wafer supplier in semiconductor industry *WSEAS Trans. Inf. Sci. Appl.,* vol. 6, no. 5, pp. 756–767
[13] Cheng J H and Tang C H 2009 An application of Fuzzy Delphi and Fuzzy AHP for multi-criteria evaluation on bicycle industry supply chains *WSEAS Trans. Syst. Control*, vol. 4, no. 1, pp. 21–34
[14] Dalkey N and Helmer O 1962 An Experimental Application of The Delphi Method To the use of experts
[15] Wright K B 2015 Researching Internet-Based Populations: Advantages and Disadvantages of Online Survey Research, Online Questionnaire Authoring Software Packages, and Web Survey Services *J. Comput. Commun.,* vol. 10, no. 3
[16] Andrews D, Nonnecke B, and Preece J 2003 Electronic Survey Methodology: A Case Study in Reaching Hard-to-Involve Internet Users *Int. J. Hum. Comput. Interact.,* vol. 16, no. 2, pp. 325–344
[17] Hsu C C and Sandford B A 2005 The Delphi technique : Making Sense of Consensus *Pract. Assesment, Res. Eval.*, vol. 7, no. 2, p. 4
[18] Hussin M R M J Z, Noh N R M, Sapar A A, and Alias N 2013 Application of Fuzzy Delphi Method in Educational Research *Design and Developmental Research*
[19] Kamarulzaman N, Jomhari N, Raus N M, and Yusoff M Z M 2015 Applying the Fuzzy Delphi Method to Analyze the user Requirement for user Centred Design Process in Order to Create Learning Applications *Indian J. Sci. Technol.*, vol. 8, no. 32
[20] Cheng C and Lin Y 2002 Evaluating the best main battle tank using fuzzy decision theory with linguistic criteria evaluation *Eur. J. Oper. Res.* 142, vol. 142, pp. 174–186
[21] Nonaka I 1994 A Dynamic Theory of Organizational Knowledge Creation *Organ. Sci.*, vol. 5, no. 1, p. 25

IOP Conf. Series: Materials Science and Engineering **645** (2019) 012011 doi:10.1088/1757-899X/645/1/012011

# Prediction of springback in the air V-bending of metallic sheets

**T Trzepiecinski[1,*] and H G Lemu[2]**
[1] Rzeszow University of Technology, Department of Materials Forming and Processing, al. Powst. Warszawy 12, 35-959 Rzeszów, Poland
[2] University of Stavanger, Department of Mechanical and Structural Engineering and Materials Science, N-4036 Stavanger, Norway

**Abstract.** Springback is a critical phenomenon in design and analysis of sheet metal forming process of metallic sheets. An accurate prediction of elastic recovery of material allows to design forming tools which take into account springback compensation. Springback is influenced by many factors including mechanical properties of material, friction conditions, temperature and geometry of bending die. In this paper, the investigations are focused on the analysis of an intelligent air bending process using an artificial neural network (ANN). The air bending experiments were carried out in a designed semi closed 90° V-shaped die. The tests were conducted on three grades of sheet metals: aluminium 1070, brass CuZn37 and deep-drawing quality steel sheet DC04. The results of experimental tests were used as a training set for back-propagation learning of a multilayer artificial network built in Statistica Neural Network program. For all materials tested, an increase of the springback coefficient is observed when the bend angle increases. The results of neural prediction are in a good agreement with the experiments. The correlation coefficient of ANN prediction to the experimental results is equal to about 0.99.

## 1. Introduction

Elastic recovery of material called springback is one of the main phenomena which influences the quality of the drawpieces formed in sheet metal forming process [1]. The accuracy and quality of sheet metal formed parts depend on the contact phenomena in sheet-tool interface and quality (wear) of the tools surface [2-5]. To control springback, there exist two main techniques: geometry-based compensation and mechanics-based reduction [6]. The geometry-based compensation can guarantee the shape accuracy of the drawpiece by performing the modifications of the tools [7]. These modifications are performed based on the measurements of the springback of a real part using digital image correlation methods or based on the results of finite element-based simulations. The other method is overbending of the sheet material. In the mechanics-based methods the sheet metal is subjected to additional tension stresses which reduce the springback. It is possible by choosing the appropriate trajectory of blankholder force variation [8].

There are many simple tests which allow the fast prediction of springback of metallic sheets. In this paper the air V-bending process is applied which allows to achieve different angles. Three main groups of parameters are influenced the springback phenomenon: geometrical, material and technological. As

* Corresponding author: tomtrz@prz.edu.pl

IOP Conf. Series: Materials Science and Engineering **645** (2019) 012011 doi:10.1088/1757-899X/645/1/012011

the bending radius increases, the springback angle increases considerably, while with an increase of thickness the springback angle decreases (Fig. 1). The springback angle decreases slightly with an increase of the sheet width. An increase of the yield stress causes an increase of the springback angle. Pre-strengthened materials, due to the greater value of plasticizing stresses, spring more than annealed materials. Since the increase in the deformation speed causes an increase in the value of the yield stress, it also causes an increase in the value of the springback angle. The effect of temperature is reversed.

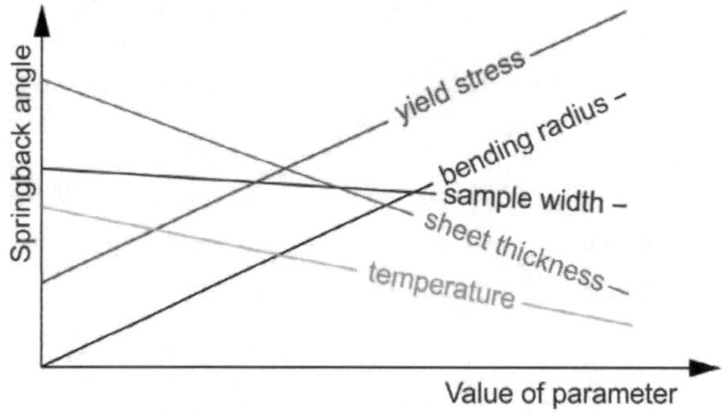


**Figure 1.** Effect of selected parameters on the springback angle.

Many studies were devoted to the development of a neural network control system for the development of an intelligent bending process. Şenol et al. [9] used ANN to predict the springback amount of stainless steel sheets through experiment based networks. It is observed that ANN can be applied effectively to determine springback in the air bending process. Jamli et al. [10] process the nonlinear elastic recovery using ANN. It was found that this approach is able to perform pattern recognition after plastic deformation. In the other work, Jamli et al. [11] processed the experimental results of monotonic loading, unloading, and reloading through a back propagation network that is able to do a direct mapping of elastically-driven change after the plastic forming. Song and Yu [12] developed a neural network for identification of residual deflection of a T-section beam after springback. The results indicate that the proposed approach could achieve an allowable straightness error. Gisario et al. [13] analysed the springback in sheet metal bending by laser-assisted bending. Multilayer perceptrons were proposed to improve the matching between the numerical and experimental data. Kazan et al. [14] developed the prediction model of springback in wipe-bending process using ANN approach. The neural networks were trained using the data of numerical simulations using finite element method (FEM). Li and Lu [15] used the radial basis function (RBF) ANN to study the springback and tensile strength in age forming of 2A97 aluminium alloy. It was found that prediction accuracy of the RBF ANN model is higher than that of the regression model. Pathak et al. [16] used the results of finite element analyses as input to ANN and gets several relevant outputs for springback compensation. Dilan et al. [17] proposed ANN-based intelligent control algorithm pipeline to eliminate the effects of variation of physical properties of sheet metals on bending. Fu et al. [18] combined three methods; FEM, genetic algorithm and ANN for optimizing the multiple step incremental air-bending of metallic sheets. Inamdar et al. [19] developed an ANN model of springback prediction based on the backpropagation of error. It was found that accuracy of predictions depended more on the number of training patterns used than on the ANN architecture.

In this paper, the ANN is used to build a model of springback prediction of three grades of materials. The experimental results of air V-bending test were used to training the ANN. The results of ANN modeling by Statistica Neural Network program are verified by results of experimental investigations.

IOP Conf. Series: Materials Science and Engineering **645** (2019) 012011 doi:10.1088/1757-899X/645/1/012011

## 2. Material

Three grades of materials were used as test specimens: aluminium 1070, brass CuZn37 and deep-drawing quality steel sheet DC04. The V-bending tests were conducted on 40 mm x 110 mm specimens cut along the rolling direction of the sheet. The thickness of specimens was 2 mm. The mechanical properties of the sheet metals (Table 1) have been determined through uniaxial tensile tests conducted according to EN ISO 6892-1:2016 standard [20]. Three samples were tested for each material. The average values of three mechanical parameters are determined using the formula:

$$X_{av} = \frac{X_0 + 2X_{45} + X_{90}}{4} \quad (1)$$

where: $X$ – the mechanical parameter, the subscripts denote the orientation of the sample with respect to the rolling direction of sheet.

The values of the strain hardening parameters K and n in Hollomon law ($\sigma = K \cdot \varepsilon^n$) are determined from the logarythmic true stress – true strain plot by linear regression.

**Table 1.** The selected mechanical properties of the tested sheet*

| Material | $E$ (GPa) | $R_{p0,2}$ (MPa) | $R_m$ (MPa) | $A_{50}$ (%) | $K$ (MPa) | $n$ |
|---|---|---|---|---|---|---|
| 1070 | 64.5 | 118.7 | 127.3 | 8.94 | 155.0 | 0.048 |
| CuZn37 | 107.5 | 285.9 | 382.7 | 25.4 | 609.0 | 0.174 |
| DC04 | 151.4 | 187.0 | 322.5 | 46.2 | 559.6 | 0.218 |

* $E$ – Young's modulus, $R_{p0,2}$, yield stress, $R_m$ – ultimate tensile stress, $A_{50}$ – elongation, $K$ – strain hardening coefficient, $n$ – strain hardening exponent

## 3. Experimental procedure

V-bending tests were carried out in experimental stand with a semi closed 90° V-die (Fig. 2). The flat specimens with dimensions of 110×40×2 mm were cut in the rolling direction. During the tests, three parameters were measured: bending force $P_B$, punch bend depth under loading $u_l$ and punch bend depth under unloading $u_u$ (Fig. 3).

**Figure 2.** The experimental setup.

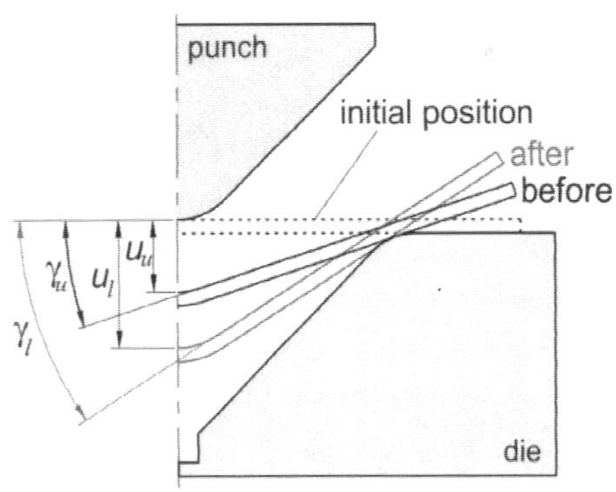


**Figure 3.** The schematic of measurement of springback.

The values of these parameters were registered by QuantumX Assistant V.1.1 program for processing the signals of both force and punch stroke transducers. Three experiments were carried out for each specimen material. The springback coefficient $K_s$ is defined as a ratio of specimen bend angle under unloading $\gamma_u$ and specimen bend angle under loading $\gamma_l$ (Fig. 3) according to the following formulae:

$$K_s = \frac{\gamma_u}{\gamma_l} \quad (2)$$

IOP Conf. Series: Materials Science and Engineering **645** (2019) 012011 doi:10.1088/1757-899X/645/1/012011

## 4. ANN Modeling

Multilayer networks named multilayer perceptrons (MLP) with a suitable number of hidden layers and neurons are theoretically sufficient to approximate any nonlinear function [21]. To build an accurate ANN, it is necessary to prepare the training set consisting of input signals and the corresponding values of output signal (springback coefficient). The following input sets of variables were assigned as input signals: yield stress of material, Young's modulus, punch bend depth under unloading and punch bend depth under loading. For the purpose of the use to keep an independent check on the progress of the training algorithm from all observations was random separated a validation set which contains 10% of all training samples.

Too large number of input variables may cause noisy data and increase the input data required for accurate training of the ANN. So, only basic inut variables were selected for the analysis. The algorithm built in *Statistica Neural Networks* program confirmed of importance of all selected input variables. The optimal network structure was found by *Intelligent Problem Solver* (*IPS*) built-in *Statitica Neural Networks.* The objective of *IPS* analysis was to find the network architecture that ensures the smallest value of standard deviation ratio (S.D. ratio) in connection with a high value of Pearson's correlation coefficient R [22]. The ANN (MLP 4 :4-7-1:1) has four neurons in input layer, eight neurons in hidden layer and one neuron in output layer The back propagation algorithm was selected to train the ANN. To prevent overlearning, the learning process was stopped when the value of verification root mean square error [22] for validation set was stopped dropping.

## 5. Results

The regression statistics for the ANN anlysed are shown in Table 2. The good quality of the MLP is confirmed by high value of correlation for both data sets. According to the Manual of Statistica Neural Networks Software [22] the very good model of the ANN is characterized by S.D. ratio value in the range of 0 and 0.1. S.D. ratio is computed as the ratio of error of standars deviation and standard deviation of output variable. The low value of S.D. ratio indicates a good ability of neural network to generalize data which were not involved in the network learning process.

**Table 2.** Regression statistics of MLP 4 :4-7-1:1 network

| Set | Data mean | Data S.D. | Error mean | Error S.D. | S.D. ratio | Correlation |
|---|---|---|---|---|---|---|
| Training | 0.8692083 | 0.1132947 | 0.0003097 | 0.0114139 | 0.1007451 | 0.9949153 |
| Verification | 0.8053333 | 0.1089694 | -0.002782 | 0.0102202 | 0.09379 | 0.9999171 |

With an increase in the value of punch bend angle under loading $\gamma_l$ the value of springback coefficient nonlinearly increases (Fig. 4). For the small values of bend angle, the material is mainly subjected to the action of elastic deformation and springback coefficient increases very fast. In the next stage of bending, the sheet is subjected to the impact of plastic deformation and its strength increases as a result of strain hardening phenomenon.

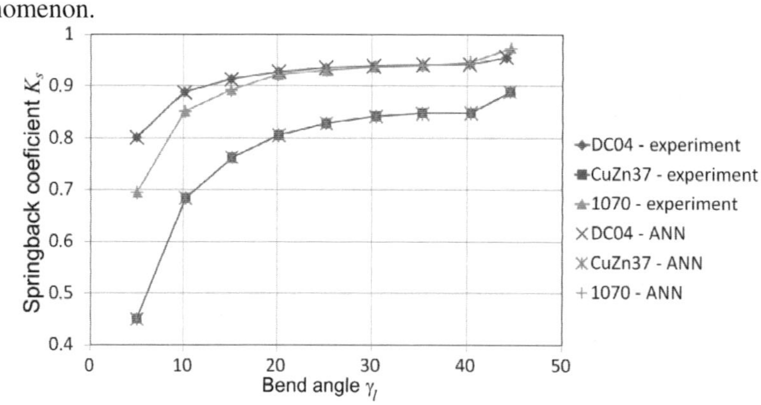


**Figure 4.** Effect of bending angle on springback coefficient.

IOP Conf. Series: Materials Science and Engineering **645** (2019) 012011 doi:10.1088/1757-899X/645/1/012011

Then the value of springback coefficient stabilizes and after bending to the angle of 45°, it slightly increases. This is related to the increased resistance of deformation after the surface of the sheet is contacted with the lateral surfaces of the die. Although the tested sheets differed significantly in terms of the value of mechanical properties, the neural network matched perfectly to the experimental data (Fig. 4). It is confirmed by the value of correlation coefficient whose value is close to 1 (Table 2).

## 6. Summary

Springback is defined as a geometrical change of the element shape after the forming process and unloading. Elastic recovery of material is one of the main problems in the forming process of drawpieces with a complex shape. By application of the artificial neural networks, it is possible to accurately predict the sheet metal springback coefficient. Furthermore, based on the neural model results, we can create the database of springback coefficient based on the experimental tests of different both grades of material and sheet thickness. After collecting a sufficiently large knowledge base, it is possible to predict the behavior of sheets with known mechanical properties without the need to carry out time-consuming experimental research.

## References

[1] Trzepiecinski T and Lemu H G 2017 *Metals-Basel* **7** 380
[2] Trzepieciński T, Bazan A and Lemu H G 2015 *Int. J. Auto. Tech.-Kor.* **16** 849
[3] Trzepieciński T and Lemu H G 2014 *Stroj. Vestn.-J. Mech. E.* **60** 51
[4] Jaworski J and Trzepiecinski T 2016 *Kovove Mater* **54** 17
[5] Jaworski J, Kluz R and Trzepiecinski T 2016 *Eksploat. Niezawodn.* **18** 271
[6] Burchitz I A 2008 *Improvement of springback prediction in sheet metal forming* (Enschede : PrintPartners Ipskamp)
[7] Lingbeek R, Huetink J, Ohnimus S, Petzoldt M and Weiher J 2005 *J. Mater. Process. Tech.* **169** 115
[8] Liu G, Lin Z, Xu W, and Bao Y 2002 *J. Mater. Process. Tech.* **120** 259
[9] Şenol Ö, Esat V and Darendeliler H 2014 *Procedia Eng.* **81** 999
[10] Jamli M R, Ariffin A K and Wahab D A 2014 *Expert Syst. Appl.* **41** 3662
[11] Jamli M R, Ariffin A K and Wahab D A 2015 *Expert Syst. Appl.* **41** 2604
[12] Song Y and Yu Z 2013 *Arch. Civ. Mech. Eng.* **13** 229
[13] Gisario A, Barletta M, Conti C and Guarino S 2011 *Opt. Laser. Eng.* **49** 1372
[14] Kazan R, Firat M and Tiryaki A E 2009 *Mater. Design* **30** 418
[15] Li H Y and Lu X C 2015 T. *Nonferr. Metal. Soc.* **25** 1043
[16] Pathak K K, Panthi S and Ramakrishnan N 2005 *Def. Sci. J.* **55** 125
[17] Dilan R A, Balkan T and Platin B E 2017 *Procedia Eng.* 207 1575
[18] Fu Z, Mo J, Chen L and Chen W 2010 *Mater. Design* **31** 267
[19] Inamdar M V, Date P P and Desai U B 2000 *J. Mater. Process. Tech* **180** 45
[20] EN ISO 6892-1 2016 *Metallic Materials — Tensile Testing — Part 1: Method of test at room temperature* (Geneva: International Organization for Standardization)
[21] Hertz J A 1991 *Introduction to the Theory of Neural Computation* (Boca Raton: CRC Press).
[22] *Manual of STATISTICA Neural Networks Software* 1988 (Tulsa: StatSoft Inc.)

IOP Conf. Series: Materials Science and Engineering **645** (2019) 012012 doi:10.1088/1757-899X/645/1/012012

# Maximum integrator gain of PID on performing position control of DC motor in the presence of Stribeck friction: Kalman conjecture approach

**T P Tumbuan[1], I P Nurprasetio[1], Indrawanto[2] and Z Abidin[1]**

[1] Mechanical Design, FMAE, Institut Teknologi Bandung, Indonesia
[2] Mechanical Production Engineering, FMAE, Institut Teknologi Bandung, Indonesia

Email: tobias.tumbuan@gmail.com

**Abstract.** The application of PID to perform position control in the presence of Stribeck friction is believed to provoke the system trapped in limit cycle oscillation. This research aims to investigates the formulation of PID that will stabilize the position control of DC motor in the presence of Stribeck friction. In order to diminish the limit cycle oscillation, the PID gain is set to meet global asymptotic stability in Lyapunov sense as Kalman conjecture is fully applied. The analysis shows that the stability of the system depends on the friction properties near Stribeck velocity. Furthermore, the proposed PID formulation is found to be easy to be applied as the maximum integrator gain applied to the system is directly proportional to the given proportional gain.

## 1. Introduction

The application of PID to control the motion of DC servo motor is frequent in manufacturing process for its simplicity. Unfortunately, while friction is introduced to the system, the system is prone to gain self-oscillation which is called as limit cycle oscillation in mathematics. The presence of limit cycle makes the system oscillate around its final position, yet the system never arrives at its final position. In this fashion, the position control system loses its accuracy.

The analysis of limit cycle due to friction for PID position control is dated back to the research led by Amstrong [1] who analyzed the existence of limit cycle via describing function and algebraic equation. The research of Amstrong indicated that the existence of limit cycle is related to the integration term of PID. Thus, Putra [2] proposed PD controller to stabilize position control of mechanical system. However, Putra found that PD controller makes the system possess steady state error. For this reason, PID cannot be easily replaced by PD controller.

In order to address the problem above, Sarot [3] devoted his research to find a good formulation of PID dealing with Stribeck friction. The observation of Sarot showed that a high derivative gain is required to reduce limit cycle oscillation. Furthermore, the computational issue of limit cycle of DC motor was investigated by Sin [4]. Marton [5] analyzed optimum PID gain via LQ design procedure. Nevertheless, Sarot, Sin, and Marton did not give any protocol to set PID gain which will remedy the limit cycle. This guideline was investigated by Jeon [6] who algebraically manipulated the trajectory of limit cycle of DC servo system under frictional atmosphere.

The attempt to stabilize the system with friction non linearity can also be done by ensuring the system that satisfies the global asymptotic stability in Lyapunov sense. One way to achieve this condition is to make the system comply with absolute stability criterion. This approach was applied by Yang [7] who employed circle criterion on stabilizing the motion of mechanical system due to friction. Moreover, Bruin [8] tried to reduce the limit cycle on the rotor dynamic system via Popov criterion. For a system of which the order is less than three, Kalman Conjecture is another alternative criterion [9]. However, the application of Kalman conjecture for a higher order system leads to a hidden limit cycle oscillation [10].

This research is the extension of the research led by Sarot [3] and Jeon [6]. It aims to find the PID rule that can reduce the limit cycle oscillation in permanent magnet DC servo motor due to Stribeck friction. In this research, such PID strategy is found by performing absolute stability criterion: Kalman conjecture. As it is built under Kalman conjecture, the PID strategy is very simple, straight forward and built in the sense of linear system. The analysis in this research shows that there is possibility to deal with Stribeck friction with only PI controller. The result of this research is very useful to the development of PID tuning based on genetic algorithm [11] since the genetic algorithm always needs a boundary of parameter which is the main concern of this research.

This manuscript is ordered in five sections. A brief introduction about DC servo motor and Stribeck friction is presented in section II. In section III, the analysis of Kalman conjecture on stabilizing the system is shown. Furthermore, a thorough numerical simulation to show the performance of the developed PID strategy is given in section IV. In section V, the conclusion of this decent research closes this manuscript.

## 2. PMDC motor and Stribeck friction

In the following, the main object of this research is explored: DC motor and Stribeck friction. The first part of this section is about the modelling of permanent magnet DC motor which is exposed by friction. Whereas, the Stribeck friction model employed on this research is reviewed in the second part.

### *2.1. Permanent magnet DC motor*

Permanent magnet DC (PMDC) servo motor is a very simple mechatronic system that changes the electrical energy into a mechanical energy in the form of rotation. The simplicity is achieved by the fact that the transient response of the electrical system is much faster than the transient response of mechanical system. Under that condition, the inductance value of DC motor coil can be neglected and the system becomes as simple as the first order system. The general block diagram of DC motor performing position control with PID algorithm is given by figure 1. In figure 1, $R_a$, $K_T$, $K_B$, $c_v$ and $J$ stand for resistance of armature ($R_a$), torque constant ($K_T$), back EMF constant ($K_B$), viscous friction ($c_v$), and rotational inertia ($J$) of the motor respectively. Furthermore, $K_P$, $K_I$ and $K_D$ represent proportional gain ($K_P$), integrator gain ($K_I$) and derivative gain ($K_D$) of the controller. In this research, it is assumed that the PMDC motor changes all electrical energy to mechanical energy. For this reason, the value of $K_T = K_B = K$.

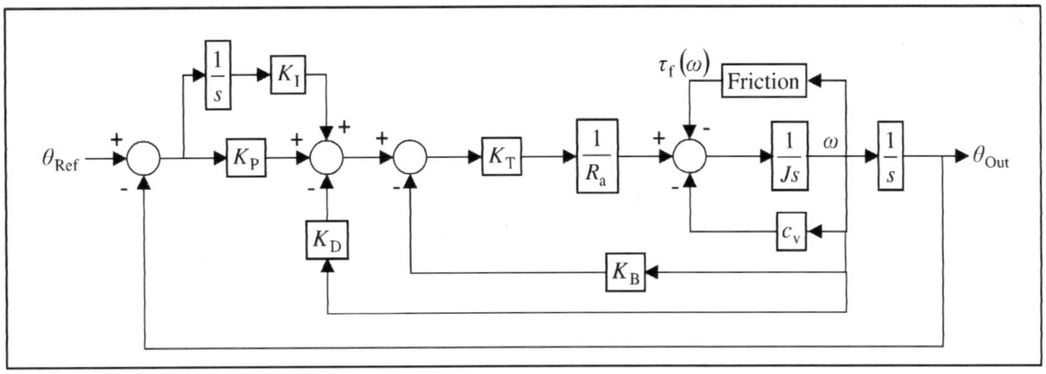


**Figure 1**. Block diagram of simplified DC motor position control with PID. The simplification is realized as the inductance of DC motor is neglected.

### *2.2. Stribeck friction*

Stribeck friction is a friction model that combines several phenomena: Coulomb friction, Stribeck effect, and viscous friction. The formula of Stribeck friction in this research is similar to the work of [3] and [4]. In this formulation, viscous friction is not taken into account due to its linear behavior. In fact, the

IOP Conf. Series: Materials Science and Engineering **645** (2019) 012012 doi:10.1088/1757-899X/645/1/012012

viscous friction is already considered in the linear system as shown in figure 1. The Stribeck friction formula which is applied on this research is shown in equation (1). In this formulation, $\tau_s$, $\tau_c$, $\omega_s$, $\varepsilon$, and $n$ represent static friction ($\tau_s$), coulomb friction ($\tau_c$), Stribeck angular velocity coefficient ($\omega_s$), angular velocity threshold ($\varepsilon$), and Stribeck coefficient ($n$) respectively.

$$\tau_f(\omega)\begin{cases} = \dfrac{\tau_s}{\varepsilon}\omega & \text{for } |\omega| < \varepsilon \\ = \left\{\tau_c + (\tau_s - \tau_c)\exp\left[-\left(\dfrac{|\omega|-\varepsilon}{\omega_s}\right)^n\right]\right\}\text{sign}(\omega) & \text{for } |\omega| \geq \varepsilon \end{cases} \tag{1}$$

The observation of equation (1) shows the pre-sliding behavior occurs when $|\omega| < \varepsilon$, whereas the sliding is developed while $|\omega| \geq \varepsilon$. In the sliding regime, the friction torque is dropped from static friction torque value to the value of Coulomb friction torque. This phenomenon is called as Stribeck effect. The severity of Stribeck effect can be represented by the minimum value of $d\tau_f/d\omega$ as shown in equation (3). Meanwhile, the nature of pre-sliding is mathematically represented by $\tau_s/\varepsilon$. These two formulations will be applied in the stability analysis via Kalman conjecture.

$$\frac{d\tau_f}{d\omega} = \begin{cases} = \dfrac{\tau_s}{\varepsilon} & \text{for } |\omega| < \varepsilon \\ = -\dfrac{n(\tau_s - \tau_c)}{\omega_s}\left(\dfrac{|\omega|-\varepsilon}{\omega_s}\right)^{n-1}\exp\left[-\left(\dfrac{|\omega|-\varepsilon}{\omega_s}\right)^n\right] & \text{for } |\omega| > \varepsilon \end{cases} \tag{2}$$

$$\left.\frac{d\tau_f}{d\omega}\right|_{\min} = -\frac{n}{\omega_s}(\tau_s - \tau_c)\left(\frac{n-1}{n}\right)^{\left(\frac{n-1}{n}\right)}\exp\left(\frac{1-n}{n}\right) \tag{3}$$

## 3. Kalman conjecture

The basic idea of Kalman conjecture is intuitive and built under the linear stability criterion. For the given system shown on figure 1, it can be stated as follow. Suppose there exists $\{k_1, k_2\} \in$ Real such that $k_1 < d\tau_f/d\omega < k_2$. The system will achieve global asymptotic stability if the system is linearly stable for any friction gain $k_{fric}$ where $k_1 < k_{fric} < k_2$. Based on this preposition, the system will be stable when G(s) is stable.

$$G(s) = \frac{\theta_{Out}}{\theta_{Ref}} = \frac{\dfrac{K}{R_a}K_P s + \dfrac{K}{R_a}K_I}{Js^3 + \left[\dfrac{K}{R_a}(K + K_D) + c_v + k_{fric}\right]s^2 + \dfrac{K}{R_a}K_P s + \dfrac{K}{R_a}K_I} \quad \text{where} \quad \left.\frac{d\tau_f}{d\omega}\right|_{\min} < k_{fric} < \frac{\tau_s}{\varepsilon} \tag{4}$$

In order to make G(s) be linearly stable, G(s) has to follow Hurwitz criterion which requires the following condition.

$$\left[\frac{K}{R_a}(K + K_D) + c_v + k_{fric}\right] > 0 \ \forall \left(\left.\frac{d\tau_f}{d\omega}\right|_{\min} < k_{fric} < \frac{\tau_s}{\varepsilon}\right) \Leftrightarrow \left(\frac{K}{R_a}(K + K_D) + c_v + \left.\frac{d\tau_f}{d\omega}\right|_{\min}\right) > 0 \tag{5}$$

$$\frac{K_I}{K_P} < \frac{1}{J}\left[\frac{K}{R_a}(K + K_D) + c_v + k_{fric}\right] > 0 \ \forall \left(\left.\frac{d\tau_f}{d\omega}\right|_{\min} < k_{fric} < \frac{\tau_s}{\varepsilon}\right) \Leftrightarrow \frac{K_I}{K_P} < \frac{1}{J}\left(\frac{K}{R_a}(K + K_D) + c_v + \left.\frac{d\tau_f}{d\omega}\right|_{\min}\right) \tag{6}$$

The elaboration of equation (5) also shows the possibility of the system to have PI controller ($K_D$=0) while maintaining the stability. PI controller is possible to be employed only when a light Stribeck phenomenon presented in the system so that equation (7) is fulfilled. In this condition, the system will be stable as long as the value of $K_I$ is less than its maximum value ($K_{I\text{-Kalamn PI}}$) which is given in equation (8).

$$\left(\frac{K^2}{R_a} + c_v + \left.\frac{d\tau_f}{d\omega}\right|_{\min}\right) > 0 \tag{7}$$

$$K_{\mathrm{I}} < K_{\text{I-Kalman PI}} = \frac{1}{J}\left(\frac{K^2}{R_{\mathrm{a}}} + c_{\mathrm{v}} + \left.\frac{\mathrm{d}\tau_{\mathrm{f}}}{\mathrm{d}\omega}\right|_{\min}\right)K_{\mathrm{P}} \tag{8}$$

Given that equation (7) is not fulfilled, the minimum $K_D$ has to be fed-back to the system denoted by $K_{D\text{-min}}$, shown in equation (9). Having $K_D > K_{D\text{-min}}$, the system will achieve global asymptotic stability if $K_I$ less than its maximum value ($K_{\text{I-Kalman PID}}$) that is shown in equation (10).

$$\left(\frac{K^2}{R_{\mathrm{a}}} + c_{\mathrm{v}} + \left.\frac{\mathrm{d}\tau_{\mathrm{f}}}{\mathrm{d}\omega}\right|_{\min}\right) < 0 \Rightarrow K_{\mathrm{D}} > K_{\mathrm{D-min}} = -\frac{R_{\mathrm{a}}}{K}\left(\left.\frac{\mathrm{d}\tau_{\mathrm{f}}}{\mathrm{d}\omega}\right|_{\min} + c_{\mathrm{v}}\right) - K \tag{9}$$

$$K_{\mathrm{I}} < K_{\text{I-Kalman PID}} = \frac{1}{J}\left(\frac{K}{R_{\mathrm{a}}}\left(K + K_{\mathrm{D}}\right) + c_{\mathrm{v}} + \left.\frac{\mathrm{d}\tau_{\mathrm{f}}}{\mathrm{d}\omega}\right|_{\min}\right)K_{\mathrm{P}} \tag{10}$$

The elaboration above shows that Kalman conjecture highlights the Stribeck effect in the stability analysis. It is pointed out as the minimum value of $d\tau_f/d\omega$ determines the possibility of PI controller as well as the minimum $K_D$ has to be applied. Moreover, the relationship between maximum $K_I$ that can be applied is proportional to $K_P$ and $K_D$. It is noted that Kalman conjecture assures the stability of the system but it does not confirm the instability of the system. The system that does not agree with Kalman conjecture is not guaranteed to be unstable. The inaccuracy to determine the real maximum $K_I$ before the system experiencing the limit cycle oscillation is the main weakness of this approach.

## 4. Illustrative example and discussion

In order to understand the application of the analysis above more in depth, an illustrative example is given. The parameter applied in this illustrative example is shown in table 1. The DC motor parameter presented in table 1 is the characteristic of UGTMEM 06 LD series. As shown in table 1, the value of $K_T$ is equal to $K_B$ as it is assumed on the modelling section. There are four cases which will be evaluated in this section. For all cases, the value of $K_P$ is set to 10, while the value of $K_I$ and asperity of Stribeck effect are varying.

Table 1. Simulation Parameter

| Parameter | Unit | Value |
|---|---|---|
| $J$ | $gm^2$ | 0.176 |
| $K_T$ | Nm/A | 0.274 |
| $K_B$ | V/(rad/s) | 0.274 |
| $R_a$ | Ω | 2.800 |
| $c_v$ | mN.m/(rad/s) | 0.216 |
| $\tau_s$ | Nm | 0.039 |
| $\omega_s$ | rad/s | 0.100 |
| $\varepsilon$ | rad/s | 0.005 |
| $n$ | - | 1.000 |

The first two cases show the application of Kalman conjecture for the system under a severe Stribeck phenomenon. This condition is remarked by the low coulomb friction torque compared to the static friction torque. The coulomb friction is assumed to be 0.02 Nm which is about one half of the static friction torque. Thus, the system needs derivative gain of which the minimum value is 1.67 per equation (9). In this illustrative example, the value of $K_D$ is set to 2.00. For the given $K_P$ and $K_D$, the maximum integrator gain of the system without considering friction nonlinearity is about 1.27e4. However, the maximum $K_I$ could be applied regarding the presence of Stribeck friction is only about 1.86e3 per equation (10). The value of integrator gain in case 1 is set to its maximum value, $K_{\text{I-Kalman PID}}$. Thus, it would be stable. Meanwhile the value of $K_{\text{I-set}}$ in case 2 is set to be 1.01e4. This $K_{\text{I-set}}$ value is eighty percent of maximum $K_I$ while friction nonlinearity is not considered. As it is much higher than $K_{\text{I-Kalman PID}}$, the system is predicted to be unstable.

The last two cases show the possibility of PI controller in stabilizing the system with light Stribeck phenomenon. In these cases, the value of coulomb friction torque is about 0.037 Nm which is only 5% lower of the static friction torque. This condition makes the PI controller is viable per equation (7). Without considering friction nonlinearity, the maximum integrator gain that could be employed is about

1.54e3. Nevertheless, as the Stribeck friction nonlinearity is considered, the maximum $K_I$ that could be applied in this system is only 3.99e2 per equation (8). The $K_{I\text{-set}}$ value in case 3 is set to $K_{I\text{-Kalman PI}}$. Hence, case 3 would be stable. Meanwhile, the integrator gain in case 4 is set to 1.23e3 which is about 80% of the maximum integrator gain without considering friction nonlinearity. Since the system in case 4 does not meet Kalman conjecture, the system is predicted to be unstable. The resume of those four cases is shown in Table 2.

**Table 2**. Simulation Parameter

| Case | Coulomb friction torque $\tau_c$ [N.m] | $K_P$ | $K_{I\text{-set}}$ | $K_D$ | Prediction | |
|---|---|---|---|---|---|---|
| | | | | | Justification | Condition |
| 1 | 0.020 | 10 | 1.86e3 | 2 | equation (10) | Stable |
| 2 | 0.020 | 10 | 1.01e4 | 2 | equation (10) | Unstable |
| 3 | 0.037 | 10 | 3.99e2 | 0 | equation (8) | Stable |
| 4 | 0.037 | 10 | 1.23e3 | 0 | equation (8) | Unstable |

The result of all cases is presented in Fig. 2. This result is generated via ODE 45 algorithm with relative error of 1e-6. The calculation is done in Scilab 6.0.0. The simulation results for case no 1, 2, 3 and 4 are orderly presented in Fig. 2 (a), (b), (c), and (d). It is shown in Fig. 2 (a) and (c) that case 1 and case 3 are stable although it is slowly decaying. This condition matches the prediction shown in Table 2. Moreover, Fig. 2 (b) and (d) show that case 2 and case 4 are trapped in the limit cycle oscillation as it violates Kalman conjecture. The amplitude of limit cycle in case 2 is about 0.006 rad, while the amplitude of limit cycle in case 4 is about 0.003 rad.

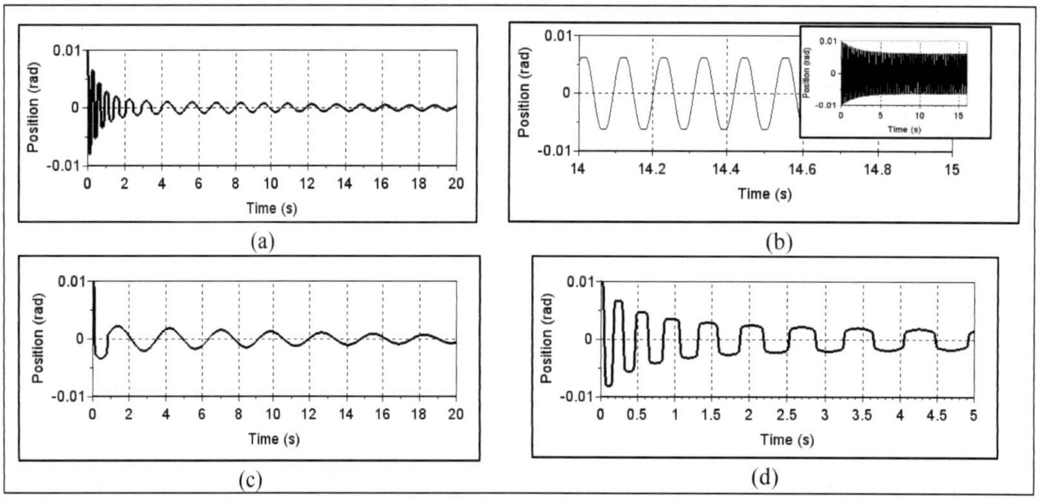


**Figure 2.** Simulation result. (a) Response of system in case 1. The system is stable as it complies with Kalman conjecture. (b) Response of system in case 2. The system is trapped in limit cycle oscillation due to the $K_I$ applied is higher than $K_{I\text{-Kalman PID}}$. (c) Response of system in case 3. The system is stable since it follows Kalman conjecture. (d) Response of system in case 4. The system is unstable since the $K_I$ employed is higher than $K_{I\text{-Kalman PI}}$.

## 5. Conclusion

In this research the PID formulation to stabilize the DC motor position control under Stribeck friction atmosphere has been well examined via Kalman conjecture. Based on the analysis, the Kalman conjecture focuses on the properties of Stribeck effect of friction. The analysis shows that the PI controller is also possible to be applied for the system with a light Stribeck phenomenon. However, for the system with severe Stribeck phenomenon, derivative gain is needed. The minimum $K_D$ for this kind of system is well formulated. Moreover, the maximum $K_I$ that can be applied to the system is proportional to $K_P$. It has to be noted that the PID formulation proposed in this research guarantees the stability of the system in the perspective of Kalman conjecture. However, the formulation is not the boundary of instability. This inaccuracy is the weakness of this research.

**Acknowledgments**
This work is supported by ITB under P3MI research grant 2017.

**References**

[1] Brian A and Bimal A 1996 PID control in the presence of static criction: a comparison of algebraic and describing function analysis Automatica **32** 5 pp 679-692
[2] Devi P, Hengk N and Nathan van de W 2007 Analysis of under compensation and overcompensation of friction in 1 DOF mechanical systems Automatica **43** pp 1387-1394
[3] Srang S, Zainal A and Hideki Y 2010 Control strategy to avoid limit cycle oscillation due to Stribeck effect in position control Proc. Reg. Conf. on Mechanical and Aerospace Technology (Indonesia: Bali)
[4] Aung M T S, Zainal A and Hideki Y 2011 Computer simulation of unstable oscillations in a class of mechatronic systems Proc. The First Int. Conf. on Interdisciplinary Research and Development (Thailand) p 91
[5] Lőrinc M and Béla L 2009 Control of mechanical systems with Stribeck friction and backlash Systems & Control Letters **58** 2 pp 141-147
[6] Soo J 2011 Integrator leakage for limit cycle suppression in servo mechanisms with stiction Proceeding of ASME 2011 Dynamic System and Control Conf. and Bath/ASME Symposium on Fluid and Power and Motion Control **1** pp 859-866
[7] Sangsik Y 1992 Stable friction compensation scheme for motion control system KSME J. **6** pp 95
[8] J C A de Bruin, A Doris, Nathan van de W, W P M H Heemels and Hengk N 2009 Control of mechanical motion systems with non-collocation of actuation and friction: a Popov criterion approach for input to state stability and set valued of nonlinerities Automatica **45** pp 405-415
[9] Taher N, Donatello M, Giacomo I, Roberto G and Murti V S 2017 Revisiting kalman and aizerman conjectures Proc. IEEE 56th Annual Conference on Decision and Control pp 523-528
[10] Leonov G A, Bragin V O and Kuznetsov N V 2010 Algorithm for constructing counter examples to the Kalman problem Doklady Mathematics **82** no. 1 pp 540 – 542
[11] Radu E P, Andrei L B, Mircea B R and Emil M P 2014 Performance analysis of torque motor systems with PID controllers tuned by bacterial foraging optimization algorithms Proc. IEEE Int. Conf. on Computational Intelligence and Virtual Environments for Measurement Systems and Applications pp 141-146

IOP Conf. Series: Materials Science and Engineering **645** (2019) 012013 doi:10.1088/1757-899X/645/1/012013

# The design of noise attenuation for gas turbine GTS 22 engine

**L R Pinandhita[1] , F Hartono[2,*] and I Prasetiyo[3]**

[1,2] Aerospace Engineering Department, Institut Teknologi Bandung, Ganeca 10, Bandung, Indonesia
[3] Engineering Physics Department, Institut Teknologi Bandung, Ganeca 10, Bandung, Indonesia

[*] Email: firman_hart@ae.itb.ac.id

**Abstract**
Nowadays, awareness of global society to the green concept where life should be in harmony with nature has been increasing. One of the ideas in the green concept is to reduce excessive sound exposures or noise. Within this context, gas turbine GTS 22 engine is one of mechanical system that generates a very high sound pressure level. Hence, it is instructive to develop noise controls for the gas turbine in order to have acceptable sound pressure level. This research is focused on the design of muffler as noise control treatment after carefully analyzing the dominant noise spectrum of the gas turbine. In this paper, numerical approach is employed to design the muffler. The results are then validated by experimental ones. Moreover, noise reduction (NR) is used to assess the muffler performances. The design of muffler in this research contains all of combination treatment: area cross section expansion chamber, perforated system (micro perforated panel), porous material, partition chamber and concentric tube. The chosen final design of muffler is combination treatment design where NR 38.77 dB is present. This NR leads to sound pressure level of 80.82 dB or 68.95 dBA exist after the muffler applied on the exhaust of gas turbine. The deviation of experimental results compared to numerical ones is 7.14% in average. Moreover, the pressure drop (ΔP) of the muffler is only 0.4042% which is acceptable to keep the mechanical performance.

## 1. Introduction

Nowadays, awareness of global society to the green concept where life should be in harmony with nature has been increasing. One of the ideas is to reduce excessive sound exposures or noise where related to the health and quality of human life. In this study, the gas turbine GTS 22 engine was used as a noise source. GTS 22 is two spool gas turbine engine which has two shafts are known as low pressure compressor and high pressure compressor. The compressor and turbines can rotate at their optimum speed which will match the local airflow inside the engine. The differences on speed of airflow can be generated the noise. Based on noise measurement, the engine SPL is about 118.5 dB at 160 Hz. Muffler are applied at the exhaust section. There are two approach used for design the muffler, they are in numerically and experimentally. As a reference for this research is uses [1-3]. A hybrid silincer is proposed by introducing MPP into the plate to elicit the sound absorption in order to compensate for the deficiency in the passband caused by the insufficient sound reflection in a certain frequency range due to weaker plate stiffness [4]. The method was used in this study is numerical.

## 2. Materials and Methods

In this research a combination treatment have been applied, below are brief explanation of each treatment,

### *2.1. Expansion chamber*

The mechanism of expansion chamber to reduce noise is uses continuity. It will appear a discontinuity cause by the changing of cross section area. The attenuation process come from the difference in acoustic pressure which makes a wavefront. A wavefront will miss-match its impedance through different cross section area or called reflected wave [5].

$$S_1 u_1 = S_2 u_2 \tag{1}$$

### *2.2. Perforated system (micro-perforated panel)*

In this research is uses MPP (micro-perforated panel) Dah-You Maa model with perforation type homogeneous pattern [6]. MPP works by using a helmholtz resonator mechanism which has two main parameters that affect its performance, namely cavity depth and perforation. MPP utilizes the viscothermal effect. MPP is useful to reduce noise on low frequency. These formulations were used to design MPP Dah-You Maa model with homogeneous pattern,

Total impedance,

$$z = r + jx_m \tag{2}$$

Acoustic resistance,

$$r = \frac{32\eta t}{\sigma \rho_0 c d^2}\left(\left[1 + \frac{k^2}{32}\right]^{\frac{1}{2}} + \frac{\sqrt{2}}{32}k\frac{d}{t}\right) \tag{3}$$

Acoustic reactance,

$$x_m = \frac{\omega t}{\sigma c}\left(1 + \left[1 + \frac{k^2}{2}\right]^{-\frac{1}{2}} + 0.85\frac{d}{t}\right) \tag{4}$$

Sound absorption coefficient,

$$\alpha = \frac{4r}{(1+r)^2 + (x_m - cot\omega D/c)^2} \tag{5}$$

Perforation ratio formula,

$$\sigma = \left(\frac{\pi}{4}\right)\left(\frac{d}{b}\right)^2 \tag{6}$$

### *2.3. Porous material*

The attenuation of porous material arises from several mechanisms. One mechanism is viscous boundary layer and the other comes from heat conduction due to the transition of isothermal, isentropic areas, fiber vibrations and friction. Delany-Bazley uses for understand the works of porous material [6,7].

Surface impedance,

$$\mathrm{z_n} = \frac{\mathrm{p}_{(0)}}{\mathrm{u}_{\mathrm{n}\,(0)}} \tag{7}$$

Normal specific impedance,

$$\mathrm{z'_n} = \frac{\mathrm{z_n}}{\rho_0 c_0} \tag{8}$$

Ratio between reflected wave and incident wave,

$$\mathrm{R} = \frac{(\mathrm{z'_n} \cos\phi - 1)}{(\mathrm{z'_n} \cos\phi + 1)} \tag{9}$$

Sound absorption coefficient,

$$\alpha_{(\phi)} = 1 - |\mathrm{R}|^2 \tag{10}$$

IOP Conf. Series: Materials Science and Engineering **645** (2019) 012013 doi:10.1088/1757-899X/645/1/012013

### *2.4. Partition chamber*
On partition chamber, the expansion chamber will separated in two or more chamber. In this research the type of partition chamber is separate expansion chamber in two. It is chosen by considering the weight factor. Then will be varied the hollow space thickness or partition wall length of partition chamber [8].

### *2.5. Concentric tube*
Concentric tube works is the same as the helmholtz resonator and micro perforated panel. Concentric tube utilizes the viscothermal effect. In MPP the ratio of perforation should not be more than 20% whereas in concentric tubes the minimum perforation ratio is 60% [6].

### *2.6. Sound Pressure Level and Noise Reduction*
In this research there are two formulations used in analyzing the design capabilities of noise attenuation, namely Sound Pressure Level (SPL) and Noise Reduction (NR) [5,8,9].

$$SPL_{overall} = 10 \log \left( 10^{\frac{SPL_1}{10}} + 10^{\frac{SPL_2}{10}} + 10^{\frac{SPL_3}{10}} + \cdots \right) \quad (11)$$

$$NR\,(dB) = SPL_{source\ overall} - SPL_{receiver\ overall} \quad (12)$$

### *2.7. Methods*
Methods that used in this research is numerical and experimental. Two approach have been used for numerical methods, there are numerical methods based on computational fluid dynamic for air flow case models and finite element method for acoustic case models [9,10]. In simulating gas flow, k-epsilon turbulence model has been used. However the design of this muffler is only viewed from the value of the acoustic performance in gas flow conditions equal to zero. Input data for numerical simulation process are the data of power turbine output gas flow from existing simulation results and SPL data or noise spectrum obtained through measurements on GTS 22 gas turbine engine directly using sound level meter (SLM) at 70000 rpm engine speed for approximately 30 seconds. The measurements gain the highest SPL of 118.5 dB at frequency 160 Hz or 119.59 dB for the overall frequency measurement were used the 1/3 octave band frequency with cut off bandwidth from 100 Hz to 800 Hz.

An experiment muffler model has been developed to validate the numerical methods. The test focused only on acoustic characteristic at zero air speed. Sample muffler in this test are made of acrylic with thickness of 2 mm as a walls and partitions material also uses rockwool with density of 60 kg/m$^3$ as porous material. The process of testing of the sample muffler uses 2 rooms, i.e. reverberation chamber and anechoic chamber (figure 1). In the reverberation chamber a sound source is placed while the sample muffler is placed in the anechoic chamber. Those rooms are separated by a partition consists of gypsum and rockwool material with a density of 60 kg/m$^3$. SLM is used to obtain SPL data on a decibel scale (dB) from an omni-directional speakers (dodecahedron) as a source of noise. In the SPL data collection process, real time SPL data from source and receiver are recorded. From the two existing data, the value of each overall SPL is then calculated. After that, the NR is also calculated [6,8].

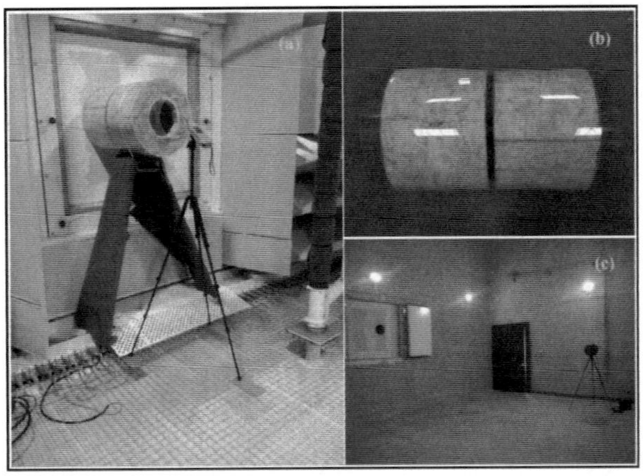


**Figure 1**. Photograph of experiment set for sample muffler. (a) Sample muffler attachment and sound level meter receiver positioning in anechoic chamber, (b) Sample muffler and (c) Reverberation chamber and position of omni-directional speaker as source

## 3. Results and Discussion

The calculation result of NR for the expansion chamber is shown in Figure 2a. For numerical simulation process the expansion chamber is varied based on the cross section ratio between the inlet muffler area and the expansion chamber area. The ratio expansion chamber starts from 1 – 3 with an interval of 0.25. In Figure 2a there are ratio of cross section expansion chamber that give a negative value NR. This negative value means that there is a reflection wave effect from sound pressure instead of absorbing the acoustic pressure from the sound source. This result agrees with the theory of the expansion chamber that the wider the area of the cross section, the lower value of the acoustic pressure. Hence on designing expansion chamber area of the muffler, it is not only by choose the highest NR value but also seen that for some frequencies the SPL values are not decrease of (attenuated). If in a certain frequency the SPL value has not been attenuated, this means the sound wave coming in is experiencing excessive reflection, so that the SPL value is not reduced but constant or even increased. From the simulation results, it was decided to choose a cross section area with a ratio of 3.

Figure 2b is the calculation results of NR for porous material. For numerical simulation of porous material, the density and thickness are varied. The density variations are 40, 60, 80, 100 and 120 kg/m$^3$ and the thicknesses variation are 25, 50 and 100 mm. Based on the numerical simulation results, the porous material with a thickness of 50 mm and density of 40 Kg/m$^3$ is selected as a design of noise muffler GTS 22 gas turbine. This selection is also based on other considerations such as the overall weight. The highest NR value was given by porous material with a thickness of 100 mm and a density of 40 Kg/m$^3$ but if using porous material with that thick and density it will increased the weight of the muffler. One thing that can be synthesized from the simulation results above is that the greater the density value of a porous material, the less absorption or attenuation ability. The effect that arises from increasing density is the sound pressure wave (acoustic pressure) will be more reflective or less able to be absorbed. Porous material with higher density values is good to be used as insulation rather than as an absorption material. Porous material is used to increase noise reduction capability for medium and high frequencies.

Figure 2c is the NR calculation result of the perforated system in the form of MPP. The numerical simulation process of MPP is conducted by varying on the thickness of the panel. MPP 1 design has a thickness of 1 mm whereas MPP 2 design has a thickness of 0.5 mm. According to the simulation results above, it is known that MPP design 1 gives a higher NR value compared to MPP design 2. This means that MPP design 1 has better noise reduction capability compared to MPP design 2 in the same frequency range. This is in accordance with the existing MPP theory that MPP works by using the Helmholtz resonator mechanism. At Helmholtz resonator there are several main parameters that affect its performance, namely the cavity depth and perforation. The frequency range and absorption coefficient value are not only determined by the air cavity but also determined by the parameters of perforation including the pore or hole diameter and the distance of pore and the panel thickness.

Figure 2d is the calculation result of NR for partition chamber. Numerical simulation of partition chamber is conducted on varying the thickness or length of the partitions. Partition chamber 1 design has a thickness or 10 mm partition distance (length), while partition chamber 2 design has a thickness or partition distance (length) of 30 mm. As the results of the simulations that have been done above, it is known that design partition 1 gives a higher NR value compared to partition chamber design 2. But it needs to be noted that partition chamber design 2 provides better attenuation on target frequency (160 Hz) than design 1.

Figure 2e is a NR calculation result of concentric tube. The difference between two kind design of concentric tube is the diameter of the perforation hole. Concentric tube 1 design has a 3 mm perforation hole diameter while concentric tube 2 design has a 5 mm perforation hole diameter. According to the NR calculations that have been done above, it is known that the concentric tube design 1 gives a higher NR value compared to concentric tube design 2. This means that the concentric tube 1 design has better noise reduction capability compared to the concentric tube design 2 in the same frequency range.

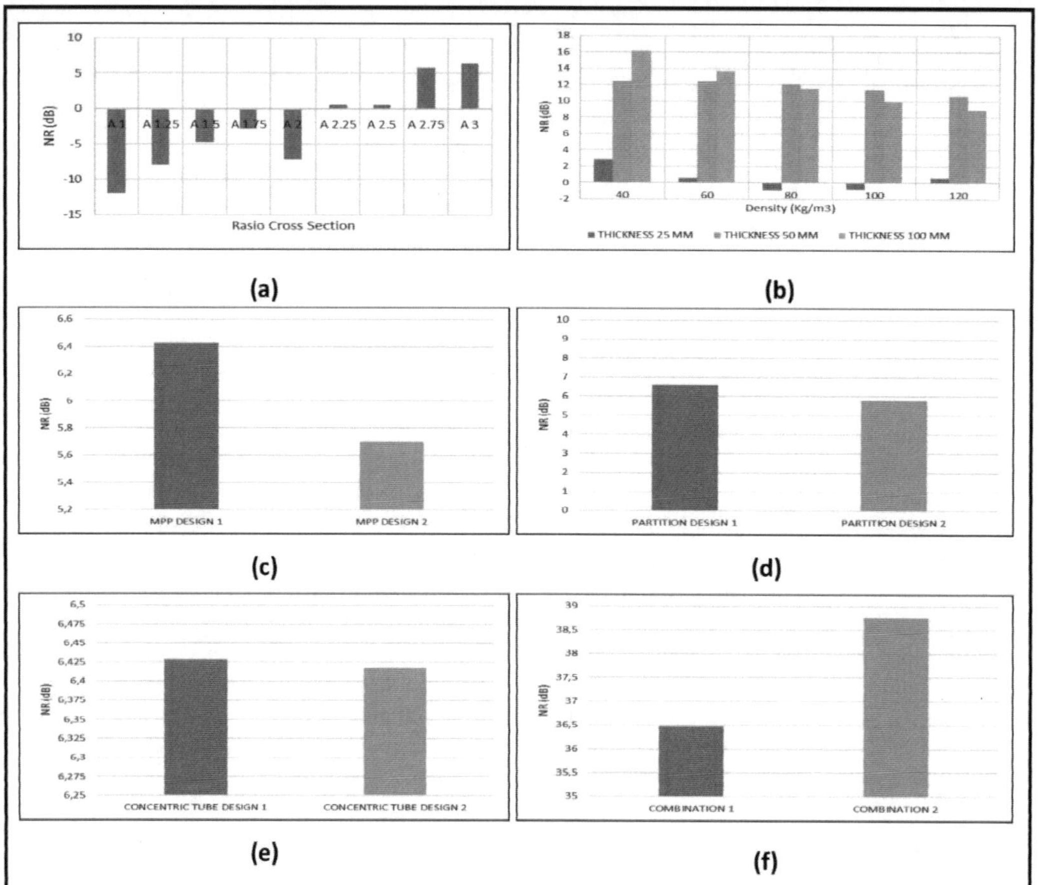


**Figure 2.** NR results of acoustic attenuation treatment in numerical simulation. (a) NR for expansion chamber, (b) NR for porous material, (c) NR for MPP, (d) NR for partition chamber, (e) NR for concentric tube and (f) NR for combination treatment

IOP Conf. Series: Materials Science and Engineering **645** (2019) 012013 doi:10.1088/1757-899X/645/1/012013

Figure 2f is the calculation result of NR for overall combination of noise reduction treatment. It obtainss a value of 38.77 dB or SPL down to 80.82 dB. When it is converted to dBA scale then it becomes 69.94 dBA. After doing the aerodynamic simulation process, the velocity of air through out the muffler is 9.1296 m/s. Hence the pressure drop value of the gas flow in the muffler can be calculated using the Bernoulli equation for incompressible flow. The value of the pressure drop ($\Delta P$) of the muffler GTS 22 gas turbine engine is 0.4042% which is acceptable in relation to maintaining the mechanical performance of the engine.

A sample muffler was used as a tool to validate the numerical result. The sample muffler in this experiment is composed of several noise reduction treatments there are expansion chamber with ratio 2, porous material 50 mm thick and density of 60 $Kg/m^3$ also partition chamber with a thickness or partition distance (length) of 30 mm. In Figure 3a there are differences in the NR values of the two methods used. This is probably because of the effect of the bulkhead between the reverberation chamber and the anechoic during the test. The inlet of the sample muffler is not parallel to the edge of the reverberation chamber and also there is a rockwool at the tip of the sample muffler inlet. Furthermore in Figure 3b, it can be seen that the trendline of the NR value for each frequency can be said has similarities. From the acquisition of NR above and the percentage level of difference or margin error between numerical method (simulation) and the test method (experiment). The percentage of the level of difference (margin error) between the numerical method (simulation) and the test method (experiment) in acoustics is 7.14% which also means that the design approach used in this study can be applied or used.

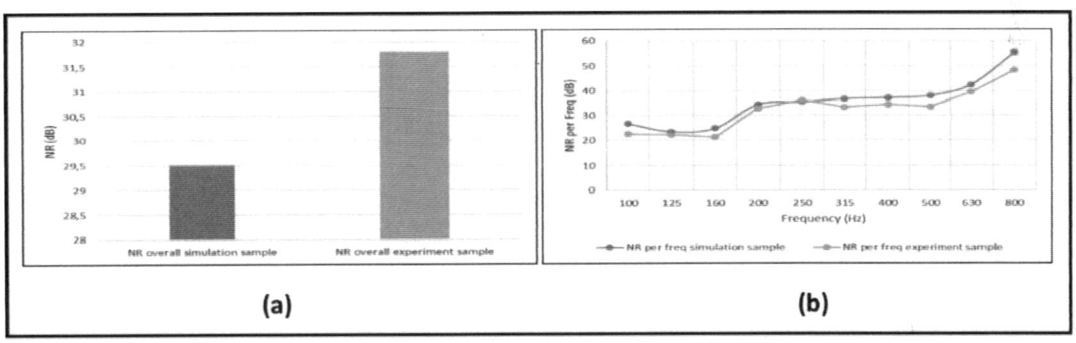


**Figure 3.** NR sample muffler. (a) $NR_{overall}$ sample muffler and (b) NR per frequency sample muffler

The muffler design for GTS 22 gas turbine engine has several noise attenuation treatments that can be used effectively:

1. Expansion chamber ratio 3 by considering the weight factor, gives NR 6.48 dB.
2. Porous material uses 100 mm thick and density of 40 $Kg/m^3$ gives NR 16.16 dB.
3. Perforation system using the MPP with Dah - You Maa model gives NR 6.43 dB.
4. Partition chamber obtaines NR 5.76 dB.
5. Concentric tube obtaines NR 6.43 dB.
6. NR of combined treatment is 38.77 dB or 80.82 dB SPL (68.94 dBA).
7. The pressure drop ($\Delta P$) is 0.4042% which can able to maintain the mechanical performance.
8. The margin error in acoustics is 7.14% which means the design approach can be applied or used.

## 4. Conclusion

The design of muffler in this research contains all of combination treatment: area cross section expansion chamber, perforated system (micro perforated panel), porous material, partition chamber and concentric tube. The chosen final design of muffler is combination treatment design where NR 38.77 dB is present. This NR leads to sound pressure level of 80.82 dB or 68.95 dBA exist after the muffler applied on the exhaust of gas turbine. The deviation of experimental results compared to numerical ones is 7.14% in

IOP Conf. Series: Materials Science and Engineering **645** (2019) 012013 doi:10.1088/1757-899X/645/1/012013

average. Moreover, the pressure drop (ΔP) of the muffler is only 0.4042% which is acceptable to keep the mechanical performance.

## Nomenclature

| | |
|---|---|
| dB | : decibel |
| dBA | : decibel A weighting |
| NR | : noise reduction (dB or dBA) |
| MPP | : micro – perforated panel |
| SLM | : sound level meter (dB SPL) |
| SPL | : sound pressure level (dB or dBA) |
| ΔP | : pressure drop (%) |
| $\eta$ | : kinematic viscosity, (kg/ms) |
| t | : panel thickness, (m) |
| $\sigma$ | : perforation ratio, ( % ) |
| $\omega$ | : angular frequency, 2Πf (rad/s) |
| $\rho_0$ | : gas density, (kg/m$^3$) |
| c | : speed of sound, (m/s) |
| d | : perforation diameter, (m) |
| k | : perforation constant |
| D | : cavity depth, (m) |
| b | : hole to hole center of perforations (m) |
| $z_n$ | : surface impedance (Pa.s/m) |
| p | : complex acoustic pressure amplitude (Pa) |
| $u_n$ | : normal acoustic particle velocity (m/s) |
| $z'_n$ | : normal specific impedance (Pa.s/m) |
| R | : ratio between reflected and incident wave |
| $\alpha$ | : sound absorption coefficient |

## References


[1] Leo L. Beranek 1996 *Acoustics* Acoustics Society of America U.S.A.

[2] M.L. Munjal 1987 *Acoustics of ducts and mufflers with application to exhaust and ventilation system design* Wiley Publication U.S.A.

[3] P.M. Morse, K. Uno Ingard 1968 *Theoritical acoustics* McGraw–Hill Book Company U.S.A.

[4] Xiaonan Wang, Y. Choy, Li Cheng 2012 *Hybrid noise control in a duct using a light micro-perforated plate* The Journal of the Acoustical Society of America. U.S.A.

[5] D.A. Bier, C. H. Hansen 2003 *Engineering noise control theory and practice* **3** Spoon Press–Taylor & Francis Group U.S.A.

[6] Prasetiyo Iwan, 2016, *Journal of mechanical Engineering and Sciences (JMES): Study on inhomogeneous perforation thick micro–perforated panel sound absorbers* UMP Publisher Malaysia

[7] Frank Fahy 2001 *Foundations of engineering acoustics* Elsevier Academic Press U.K.

[8] D.W. Herrin 13 *Design of mufflers and silincers* University of Kentucky–Departement of Mechanical Engineering U.S.A.

[9] D.P. Jena, S.N. Panigrahi 2015 *Estimating acoustic transmission loss of perforated filters using finite element method* Elsevier U.K.

[10] D.W. Pepper, J.C. Heinrich 2006 *The finite element method basic concepts and applications* **2** Taylor & Francis Group U.S.A.

IOP Conf. Series: Materials Science and Engineering **645** (2019) 012014 doi:10.1088/1757-899X/645/1/012014

# Hydrodynamic analysis of Remotely Operated Vehicle (ROV) Observation Class using CFD

**D Satria[1,*], R Wiryadinata[2], D P A Esiswitoyo[1], M I Adji[1], I Rosyadi[1], E Listijorini[1], Sunardi[1]**

[1]Mechanical Engineering Department, Universitas Sultan Ageng Tirtayasa, Indonesia
[2]Electrical Engineering Department, Universitas Sultan Ageng Tirtayasa, Indonesia

[*] Email: dhimas@untirta.ac.id

**Abstract.** Remotely Operated Vehicle (ROV) is a small marine craft used for underwater observation. Hence, the analysis of hydrodynamic characteristics is important. Hydrodynamic characteristics can be obtained by fluid dynamic visualization and simulation of *Computational Fluid Dynamic* (CFD). In this paper, ROV Hydrodynamic characteristics begin by forming a three-dimensional (3D) ROV model, forming link elements (mesh) from the 3D ROV model, and forming the computational domain used for CFD simulations. CFD simulations were conducted by analyzing the fluid flow angle variations that represent motion of the ROV, namely translational motion and rotational motion towards the vertical axis (yaw). The simulation results obtained were compared with the previous ROV design that had been developed, so as to obtain an evaluation of the results of the design optimization. The results showed an increase in the hydrodynamic characteristics of the optimal ROV design indicated by the value of drag, pressure distribution, and fluid flow contour.

## 1. Introduction

The cheapest and easiest water observation method to obtained information about water is by using observation equipment, namely remotely operated vehicle (ROV). In this case, there are several studies to develop ROV as observation equipment with various shapes. Every shapes and design has different hydrodynamic characteristic. The easiest way to obtain hydrodynamic characteristic is by simulation by using software of Computational Fluid Dynamic (CFD) compare to experiment [1,2]. In study of Amjith et al [3], analysis of hydrodynamics characteristic with the consideration of voltage and pressure contour was done by using software of ANSYS. Yue et al [4] used represent of 3 degree freedom of movement from observation to obtain drag coefficient with the software of ANSYS and in term of the deviation value through simulation by manual calculation.

Hydrodynamic analysis in this research aims to obtain hydrodynamic characteristic from developed optimum design of ROV [5], and considered from the comparison both design. Because good geometry can minimize energy required for ROV to move in fluid in determined speed [6]. As a note, the new design of ROV model used in this study (Figure 1) is the result of conceptualization of ROV design of previous design[7]. Therefore it is expected that the hydrodynamic characteristic of optimum design give better character compare to previous design.

**Figure 1.** New design of ROV

## 2. Materials and Methods

Hydrodynamic characteristics can be obtained by fluid dynamic visualization and simulation of *Computational Fluid Dynamic* (CFD). In this paper, ROV Hydrodynamic characteristics begin by forming a three-dimensional (3D) ROV model, forming link elements (mesh) from the 3D ROV model, and forming the computational domain used for CFD simulations.

### *2.1. Simulation preparation*

Simulation was conducted by forming 3D model beforehand from the concept of ROV body. The case of fluid flow in this study was sea wave of Sunda strait, which has the character of incompressible [8,9], open channel, and turbulent because ROV works in open sea water in which the flow is not uniform [8,10]. The solution toward turbulent flow, flow simulation using $k - \varepsilon$ model, which is two equations consist of turbulent kinetic energy and dissipation ratio [11].

The general simulation setting consists of the value of entering the simulation parameters. The wall condition used is the adiabatic wall. Meanwhile the initial and ambient conditions used are pressure and potential pressure P = 999600 Pa, and T = 305.5 K. Assuming that the ROV sinks to a depth of 100 m and move with an average speed $v = 2\ {}^{m}/_{s^2}$.

### *2.2. Link elements (mesh) and computational domain*

Mesh in simulation computation is fully controlled by flow simulation automatically, by simplifying 3D model into mesh model Figure 2 shows the fluid flow *Computational Domain* with the size of the computational field proportional to the size of the ROV. In the computational area it has been shown the fields of in flow, openings, and walls.

This study uses a fixed mesh method, in which it is assumed that the fluid moves at the speed at which the ROV moves and the ROV model is silent (static). Figure 3 shows representation of the motion of the ROV was conducted by changing the slope of the ROV to the direction of fluid flow. Two simulation conditions were used as representations of ROV motion, those are:

- condition 1, the fluid flow rate is parallel to the object that represents heave motion.
- condition 2, variation of the ROV slope towards the direction of the fluid flow rate with a slope variation from 10 ° to 80 ° which represents the diving motion.

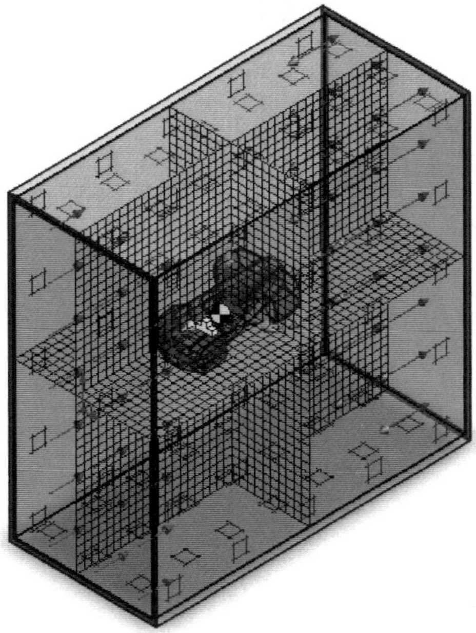

**Figure 2.** Computational domain and mesh

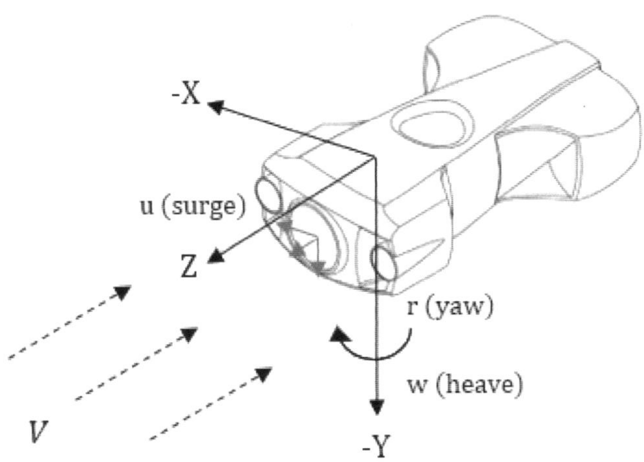


**Figure 3.** Flow simulation coordinate system

*2.3. Simulation objective*
The objective of the simulation is to obtain the value of the force and moment generated due to the fluid flow experienced by the ROV body. Tension and moment values are generated from the three static cases of the link element in the three coordinate axes; style values for drag force and lift force, and moment values. The drag coefficient value is also obtained from the objective equation by entering the drag coefficient equation into the simulation.

$$C_d = \frac{F_D}{\rho\, A \frac{V^2}{2}} \tag{1}$$

## 3. Result and Discussion

By carrying out objective resolution simulations with mesh and computational fields, and general simulation settings. Simulation forms the results in the form of flow visualization plots and objective plots in the form of values for each iteration. Flow simulation uses the number of iterations as 570-1600 and interval analysis as 120 which is determined automatically by Flow Simulation with consideration of the model geometry and general simulation settings. Mesh and the number of iterations are formed automatically with the algorithm to obtain results that fot to the best abilities and results.

### *3.1. Simulation convergence*

The flow simulation was conducted by using the average formulation of the Navier-Stokes equation so that it is not determined how the software defines convergence. However, the determination of convergence data from the simulation results was determined by variable computing on initialization and boundary conditions, so that the results obtained do not change significantly. From Figure 4, a good residual value is below 1, where the moment on the z-axis and the force on the y-axis have the lowest convergence.

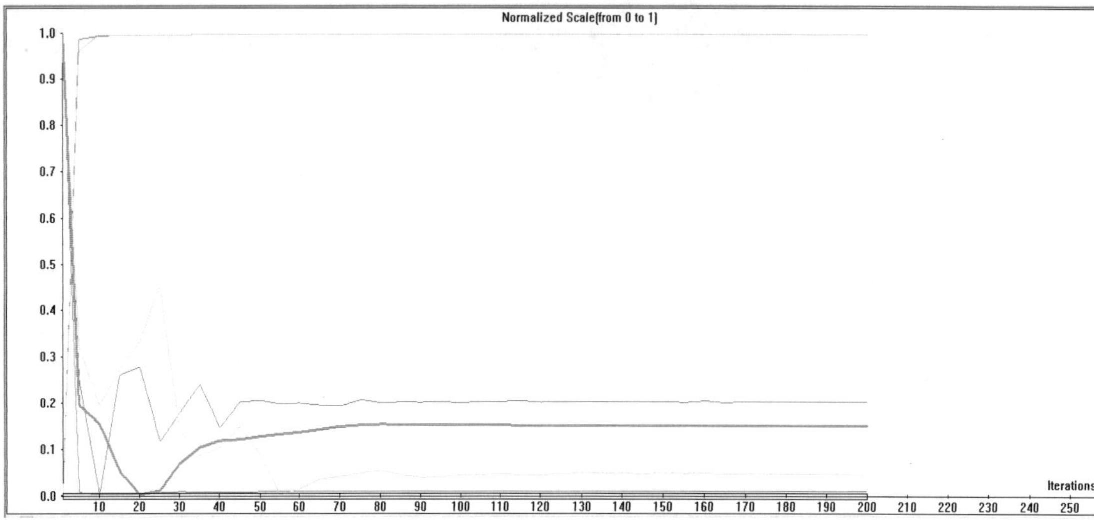


**Figure 4.** Residual value simulation

### *3.2. Horizontal motion simulation*

ROV moves relatively to the speed of fluid flow and static ROV to the speed of constant fluid flow. Figure 4 is the visualization of fluid flow in speed parameters. The fluid flow profile forms a stagnation area on the front of the ROV. As the result of the stagnation area on the front of the ROV, the speed parameter value has expanded to 2.5m/s. The flow tends to degrade from the stagnation area (forebody) to the back of the body with a good profile differentiation. The boundary layer is formed and limits between the surface of the object with an area where viscosity is ignored, has a smooth gradient. The gradient gives a low amount of friction drag. In addition, visualization of flow profiles shows good flow separation, which increases the hydrodynamics of the ROV body.

Variables obtained from the simulation of horizontal translational motion represent the value of the drag force and lift force. The value of drag force and lift force obtained were: -74.15 N and -12911.47 N. The drag force obtained tends to form friction drag because the geometry and wake phenomena resulting from the flow separation tend to be low. The fixed mesh method defines a vector where the fluid v moves in the direction of the negative z-axis with the assumption that it is normalized as a vector to the positive z-axis, which represents the thrust of the thruster. Thus, the value of the drag force and lift force obtained was negative, which shows the force opposite to the load (load; weight and thrust force) of the ROV.

### *3.3. Diving motion simulation*

ROV slope variation illustrates the motion of diving from the ROV to the critical point of diving capability (90 °), thus the ROV attack angle is varied by changing the slope of the ROV to the fluid flow rate. Other simulation settings use the same values as horizontal translational motion simulations. Figure 5 is a visualization of the fluid flow profile to changes in the slope angle of the ROV. Fluid flow separation due to the asymmetrical ratio of contact surface area and the resulting Eddy motion exacerbates the effect of wake phenomenon so that the drag value increases.

This phenomenon is proven by the force obtained from the simulation of the ROV slope variation and it is displayed in the linear graph that it is gradually increasing to its critical point. Figure 6 shows a graph of the correlation between forces on the z-axis and ROV slope. The graph shows the linearity of the data (shown in the trend line), where the greater the ROV slope angle, the greater the force produced on the z-axis. Linearity of force relationship data with slope is also obtained on the other two axes.

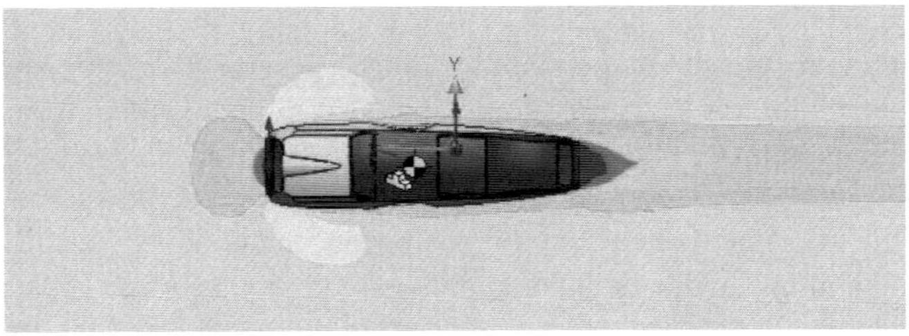

**Figure 5.** Flow profile of speed profile

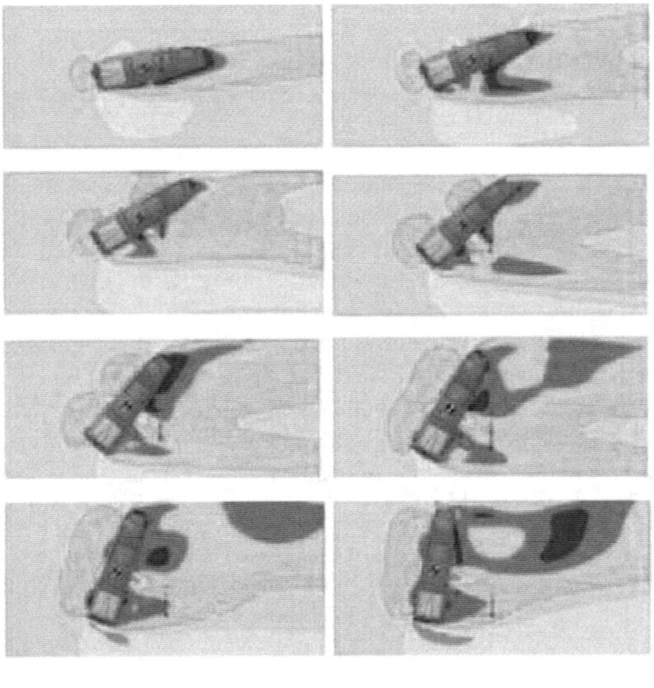

**Figure 6.** Variation of slope, respectively; 10°, 20°, 30°, 40°, 50°, 60°, 70°, and 80°

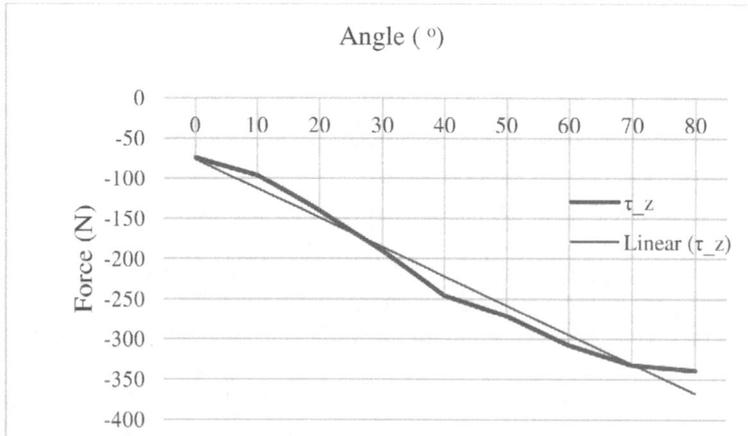


**Figure 7.** Linearity of drag force value

Figure 7 shows that the greater the slope, the more force the ROV needs to dive. The graph shows negative linear data because the direction of flow velocity is assumed to be normal to the negative z-axis, so the direction of flow velocity is positive. The force on the z-axis due to the load of the fluid flow results in a design resistance that is opposite of the direction of fluid flow Design resistance in this context is the force of the simulation results on the z-axis and y-axis. These conditions in the context of a vector are defined in the opposite direction, or negative. The acquired resistance cannot represent the real dive force. The diving resistance force is obtained from the resultant of the two resistance vectors, to the slope.

The drag force relative to the direction of velocity and lift force is perpendicular to the direction of flow velocity. Resistance on the z-axis and the -y axis is the gait and lift force on the flow direction if it is assumed that the ROV remains in its position. To get the thrust when diving, the consideration of the principle of Archimedes is applied. If it is assumed that the ROV is diving at its slope, the ROV thrust required is the resultant force of resistance to the slope, hydrostatic force on the bottom surface of the ROV (force on the y-axis), and the specific gravity of the ROV and fluid density.

### *3.4. Comparison of numerical variable design*

Numerical hydrodynamic variables as comparative aspects are measured values. Therefore this section compares the values obtained from the simulation results. The first aspect is drag force. As previously mentioned, the two designs use a different form. This difference affects the tendency of the drag force. Old designs tend to form drag and optimal designs tend to form friction drag. The drag force of the old design obtained from the simulation was 88.13 N while the drag force of the optimal design is 74.15 N. There is a large increase in resistance or the drag force in the optimal design as 18.8%. From the drag force can be calculated the drag coefficient of the two designs.

Variables that need to be determined to obtain the drag coefficient are the contact surface area. Contact surface area is measured from the 3D design model. The contact surface area for the old design was $A_l = 0.028\ m^2$ and the optimal design of $A_o = 0.040\ m^2$. Then the drag coefficient was obtained as follow:

$$C_{d_l} = \frac{2 \times 88{,}13\ N}{1020\ {}^{kg}/_{m^3} \times (2\ {}^{m}/_{s})^2 \times 0{,}070} = 1{,}60$$

$$C_{d_o} = \frac{2 \times 74{,}15\ N}{1020\ {}^{kg}/_{m^3} \times (2\ {}^{m}/_{s})^2 \times 0{,}040} = 0{,}91$$

The calculation of the drag coefficient theory of both designs shows a significant difference between the old design and the optimal design. The old design has a drag coefficient value of $C_{d_l} = 1{,}60$ and the optimal design has a drag coefficient of $C_{d_o} = 0{,}91$. The increase in drag coefficient was 43% from the optimization results. Comparison of the free variable drag coefficient between the old design and the optimal design was 1:2. The greater the ratio of the independent variable (drag force versus the area of contact area) the better the value of drag coefficient would be. In addition to the comparison between the independent variables, the old design fluid flow separation forms a fairly large wake area with a pressure reduction value which was 612.10 Pa from an environmental dynamic pressure as 2448.20 Pa.

## 4. Conclusion

Hydrodynamic analysis in this research aims to obtain hydrodynamic characteristic from developed optimum design of ROV. The result shows an increase in the hydrodynamic characteristics of the optimal ROV design indicated by the value of drag, pressure distribution, and fluid flow contour. The old design has a drag coefficient value of $C_{d_l} = 1{,}60$ and the optimal design has a drag coefficient of $C_{d_o} = 0{,}91$. The increase in drag coefficient from the optimization result is 43% .

**Acknowledgments**
The authors would like to send the gratitude to the Directorate of Research and Community Service (DRPM) of DIKTI who has funded this research in the 2018.

## References

[1] J. Ramirez-Macias, P. Brongers, S. Rua, and R. . Vasquez, "Hydrodynamic modelling for the remotely operated vehicle Visor3 using CFD," *IFAC-PapersOnLine 49-23*, pp. 187–192, 2016.
[2] H. Zhang, "Using CFD Software to Calculate Hydrodynamic Coefficients," *Mar. Sci. Appl.*, vol. 9, pp. 149–155, 2010.
[3] K. Amjith, V. Arundas, M. Harikrishnan, J. Sebastian, and V. Vineeth, "Underwater Search and Rescue Device," *Int. J. Innov. Res. Sci. Technol.*, vol. 1, no. 11, pp. 481–488, 2015.
[4] C. Yue, "ANSYS FLUENT-Based Modeling and Hydrodynamic Analysis for a Spherical Underwater Robot," *IEEE Int. Conf. Mechatronics Autom.*, vol. 13, pp. 1577–1581, 2013.
[5] R. Wiryadinata, A. S. Nurliany, I. Muttakin, and T. Firmansyah, "Design of a Low Cost Remotely Operated Vehicle with 3 DoF Navigation," *Bull. Electr. Eng. Informatics*, vol. 6, no. 1, pp. 13–23, 2017.
[6] D. L. Paster, "Importance of Hydrodynamic Considerations for Underwater Vehicle Design," *IEEE*, pp. 1413–1422, 1986.
[7] D. Satria, D. P. A. L. Esiswitoyo, N. K. Caturwati, E. Listijorini, and R. Lusiani, "Body Design Concept of Remotely Operated Vehicle (ROV) of Observation Class with the Method of Concept Screening and Concept Scoring," in *MATEC Web Conf. Volume 218, 02009, The 1st International Conference on Industrial, Electrical and Electronics (ICIEE 2018)*, 2018.
[8] Y. . Cengel, *Fundamental of Fluid Mechanics*. New York: McGraw-Hill, 2006.
[9] D. Satria, Haryadi, R. Austin, and B. Kurniawan, "Design of drying chamber and biomass furnace for sun-biomass hybrid rice-drying machine," *AIP Conf. Proc.*, vol. 1717, 2016.
[10] T. I. Fossen, *Handbook of Marine Craft Hydrodynamics and Motion Control*. Trondheim: John Wiley & Sons Publication, 2011.
[11] H. . Versteeg, *An Introduction to Computational Fluid Dynamics: The Finite Volume Method*. Pearson Education Limited, 2007.

IOP Conf. Series: Materials Science and Engineering **645** (2019) 012015 doi:10.1088/1757-899X/645/1/012015

# Development Testing Method and Analysis Static Thrust for Propeller Based Propulsion

**M Khasyofi[1,*] and F Hartono[2]**

[1]Jalan Ciung Wanara No 4 Bandung
[2]Labtek II FTMD, Institut Teknologi Bandung, Jl. Ganesha 10 Bandung, Indonesia **

[*]Email: khasyofi@gmail.com

**Abstract**. Propulsion system development involved aircraft manufacturer and engine or propeller manufacturer to perform test and evaluation. The main of this research was propulsion testing for unmanned aerial vehicle by evaluating thrust in static condition. Static thrust testing was conducted by force measurement using load cell and data acquisition systems. Thrust measurement result was compared with analytical method and computer fluid dynamic simulation. Analytical method in this paper used vortex blade element theory to calculated static thrust meanwhile Computational fluid dynamic was running both 2 dimension and 3 dimension simulation. Vortex blade element theory analysis combined with computational flight dynamic simulation for propeller aerofoil at 70% location. Test result compared with analytical and CFD calculation using non dimensional parameter. Analytical calculation deviated 5.08% from experiment meanwhile CFD calculation give 13.56% but still in error tolerances. Error tolerances considered from friction force, sensor accuracy and air density correction with total 6.33 kgf.

## 1. Introduction

Propulsion system is the main system to generate thrust and convert chemical or electric energy to mechanical energy. Mechanical energy is used to change air momentum and generate thrust. The principle on the change of momentum and thrust was explained by Newton law [1]. Aircraft propulsion divided into several classification based on operational and flight regime. Aircraft propulsion system divided into propeller based, air breathing and rocket [2]. Some aircraft propulsion system generate change of momentum in different way. The example is rocket propulsion that produce thrust using fuel and oxidizer reaction without burning reaction with air meanwhile air breathing propulsion using oxygen from air to burn fuel in combustion chamber.

The other way to generate thrust is using propeller. Propeller generated thrust by changing air momentum after passing propeller disk. Change of momentum from propeller as function of axial velocity before and after passing propeller. Propeller analysis explained by several theory such axial momentum theory, general momentum theory, blade element theory and vortex theory [1]. Momentum theory is classical theory for propeller analysis. Propeller is replaced by thin disk model and airflow around propeller described by stream tube with constant pressure at each section. In the other hand blade element theory model propeller from each cross section as airfoil. Each airfoil produced lift and drag and convert to torque and thrust using trigonometry principle. The integration of thrust and torque at each cross section was thrust and torque of the propeller.

The objective of this research was to develop static thrust calculation method and validate with test data. Test methodology used load cell and data acquisition system to record thrust and propeller revolution. Test execution was held at PT Dirgantara Indonesia facilities in Bandung.

## 2. Static Thrust Experiment

### *2.1. Static Thrust Measurement*

Static thrust model is different from dynamic thrust or advanced thrust. Calculation model for static thrust estimation is function of rotational speed, meanwhile dynamic thrust is influenced by airspeed [3].

$$Ts = 1.283.\,10^{-12}\,RPM^2\,D^4\,\rho\,Kt \tag{1}$$

Static thrust measurement was performed at Indonesian aerospace using load cell data acquisition system as illustrated in Figure 1. The aircraft was connected to load cell using wire or belt [4]. Data acquisition system recorded thrust data from several throttle setting from idle to maximum rpm. Figures 1 showed test method and illustration for static thrust measurement [5].

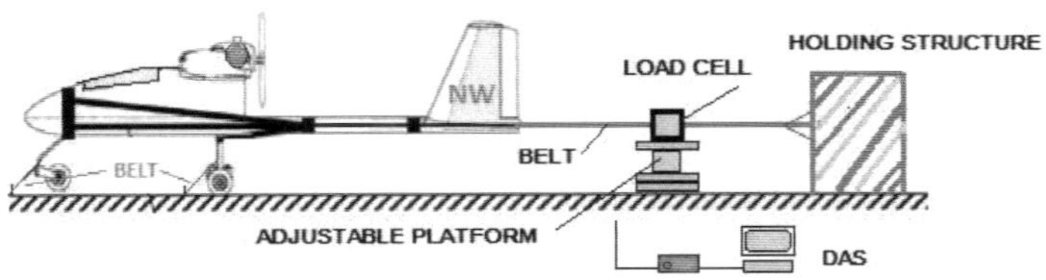


**Figure 1.**Test component and illustration static thrust measurement

Test procedures and preparation for this measurement referred to test document released by Indonesian aerospace. Before engine start, aircraft position have to be located at ground level condition. Landing gear structure and belt that was connected to load cell have to be installed completely. Emergency panel switch for engine shutdown have to be installed for safety.

Test execution was held several times at Bandung and Pangandaran. Ground test mostly held in Bandung and flight test was perform in Pangandaran. Table below showed the example of test data was taken at April 2015, as on Table 1.

**Table 1.** The example of static thrust data

| Propeller Revolution (rpm) | Thrust (Kgf) |
|---|---|
| 2000 | 1.7 |
| 3700 | 11.8 |
| 5000 | 22.1 |
| 5600 | 28.3 |
| 6000 | 33.6 |

Static thrust data from this measurement is corrected by friction from aircraft tire and ground. Friction force is calculated by pulled aircraft using dynamometer. Maximum force recorded before aircraft move from static condition was identified as rolling friction force.

Error analysis from static test is considered from instrument accuracy, friction forces and air density correction. Instrument accuracy is about 0.25 kilogram forces based on load cell specification. Friction force is about 4.5 kilogram forces and density correction is 1.58 kilogram forces. Total error is about 6.33 kilogram forces. Random error was not considered because of amount of data was limited. Some data at low rpm didn't record during test. It influenced data quality and test result.

IOP Conf. Series: Materials Science and Engineering **645** (2019) 012015 doi:10.1088/1757-899X/645/1/012015

## 3. Propeller Modelling and Simulation

### *3.1. Propeller CAD Model*

Propeller model for simulation reconstructed using reverse engineering method. Propeller model was created using mould and manual measurement in order to get propeller dimension and cross section profile. Propeller cross section generated by propeller model or propeller dummy made from polyurethane material as shown in figure 2-a below.

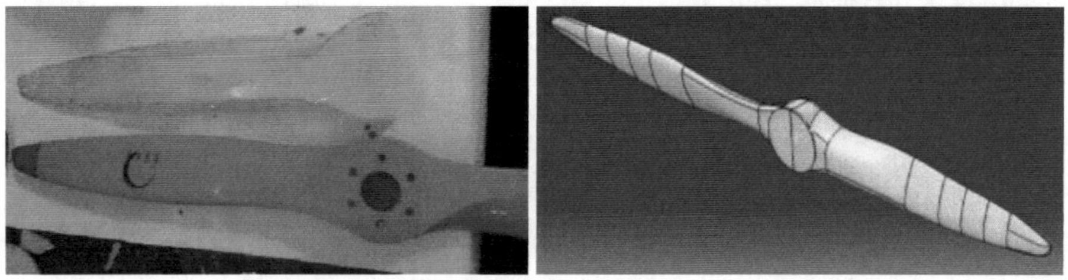

(a) Propeller model from polyurethane (b) Propeller CAD model

**Figure 2a-b**. Propeller model and CAD model

Propeller geometry such as pitch and chord are measured by calipers. 20 cross sections are selected to measured and created propeller geometry and generated CAD (Computer Aided Design) model. CAD model as shown in figures 2.a-b is reconstructed using commercial CAD software. For computational fluid dynamic simulation, CAD model would be modified in order to create mesh or topology. Some geometry simplification to create CAD model such as simplified and reduces number of section in CAD software.

### *3.2. Computational Fluid Dynamic Simulation*

Propeller simulations both 2 dimension and 3 dimension are conducted using commercial CFD software. Two-dimension simulation purpose was to generate aerofoil characteristic data (Cl and Cd) and it used for analytical calculation [6]. Three-dimension simulation directly calculate static thrust in several rotation speed with setting engine form idle to maximum power (2000-6000 rpm).

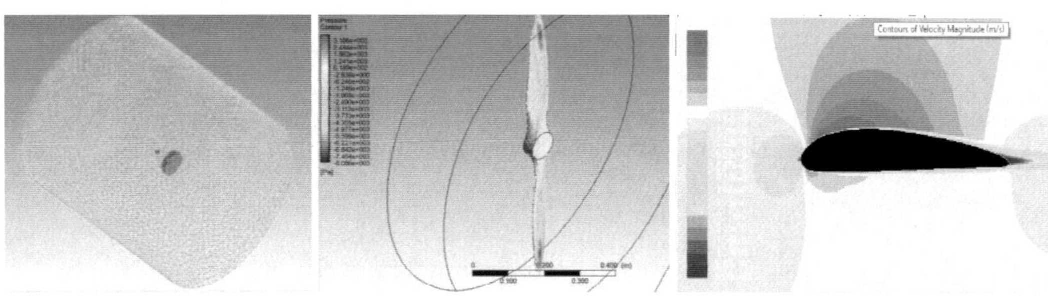

(a) Mesh and domain (b) 3D simulation (c) 2 D simulation

**Figure 3**. CFD analysis from 3D and 2D simulation result

2D simulation is calculated at airspeed 0.37 M and Reynolds number (Re) 500.000. This condition is based on propeller rotation at 4000 rpm and airfoil section at 70% from hub considered as average condition during test. 3D simulation performs at ISA and steady state with "opening" boundary condition. Meshing for this simulation used unstructured mesh with 2 fluid domain (rotary and stationary). Total mesh that used for rotary and stationary domain was about 1.6 millions. Rotary domain shaped cylindrical with diameter 1.2 of propeller and thickness 0.4 diameter of propeller [7]. Stationary domain diameter was 11 and thickness 7 times of propeller diameter. Stationary domain and rotary domain connected by "interface". Rotary domain simulated using frozen rotor approximation with variation of 2000, 3000, 4000, 5000 and 6000 rpm.

IOP Conf. Series: Materials Science and Engineering **645** (2019) 012015 doi:10.1088/1757-899X/645/1/012015

### *3.3. Analytical Calculation Using Vortex Blade Element Method*

Analytical calculation using vortex-blade element theory is combination from blade element theory and momentum theory method [1]. Calculation process was iteration process using value of induced velocity and free stream velocity. Accuracy of this method determined by aerofoil characteristic. Aerofoil characteristic for this calculation was taken from CFD simulation at 70% from hub. Assumption from this calculation that aerofoil at 70% hub represented aerofoil characteristic of propeller. Even though this condition deviate from real condition but this approximation needed to simplified calculation. The result would be compared with experiment and CFD simulation.

### *3.4. Result and Discussion*

The result of static thrust calculation are shown at Figure 3 and Table 2. Dot and blue line (experiment corrected) is experiment data after correction with friction and conversion to standard sea level [8]. The red line is CFD simulation and blue line is analytical calculation. As we can see, both simulation and analytical calculation showed the same tendency or trend with experiment result. The value from both calculation was in error range of experiment.

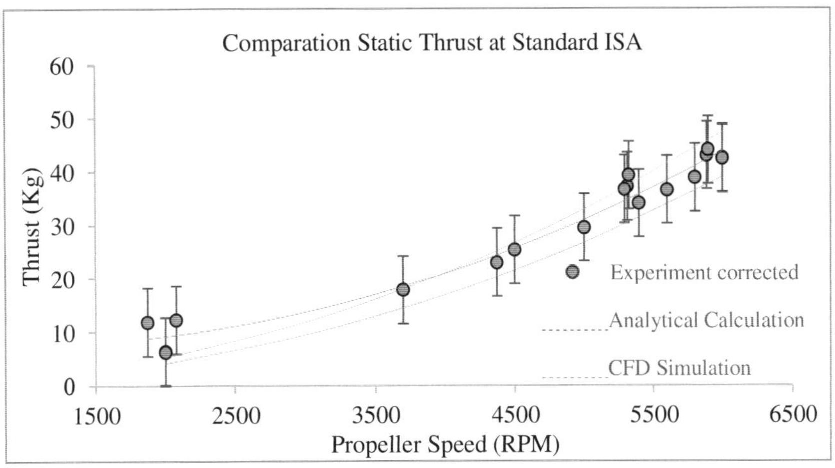


**Figure 4**.Comparation static thrust from analytic

Non dimensional comparison showed analytical calculation gave accuracy 5.08% meanwhile computational fluid dynamic simulation gave accuracy -13.56%. Static thrust calculation from CFD simulation was lower than analytic calculation and experiment. Even though this differences was in error range, thrust at low speed test data was not too close with experiment. If we see in figure 4, there was no thrust data between 2000-3500 rpm. Thrust data at low rpm should be added to validate static thrust calculation.

**Table 2.** Result and comparison static thrust non-dimensional parameter

| Calculation Method | Static Thrust Coefficient (Kt) | Deviation from Experiment (%) |
|---|---|---|
| Experiment | 0.59 | 0 |
| Analytic | 0.62 | 5.08 |
| CFD | 0.51 | -13.56 |

IOP Conf. Series: Materials Science and Engineering **645** (2019) 012015 doi:10.1088/1757-899X/645/1/012015

## 4. Conclusion

Static thrust determination using experiment, analytic and computer fluid dynamic simulation method could be used for aircraft development. The result from all method gave the same tendency and deviate in range of error margin. Error margin was 6.33 kgf, it was considered from instrument error, friction forces and atmosphere condition. Friction force measured using dynamometer by pulling aircraft and record maximum force before aircraft displace from initial position. Static thrust calculated from analytical calculation and CFD simulation and it could be compared with test data and gave deviation between 5.08-13.56 %. Analytical calculation using vortex blade element method was good enough to predict static thrust using aerodynamic data from aerofoil at 70% from hub.

## References

[1] Durand William Frederick 1934 *Aerodynamic Theory* Stanford University California

[2] Mattingly Jack D 1996 *Element of Gas Turbine Propulsion* McGraw-Hill Singapore

[3] Brandt, John B; Selig Michael S. *Propeller Performance Data at Low Reynolds Numbers.* University of Illinois at Urbana-Champaign, Urbana.2011

[4] Mingtai, Chen. *Static Thrust Measurement for Propeller driven Light Aircraft.* The 2nd International Conference on Computer Application and System Modeling.2012

[5] Wilana, Binbin. *PTTA Wulung Ground Test Procedure of Propulsion System*, Engineering Data Management PT DI D532NW5001, April 2015.

[6] Nirwanto, Ricky Tonny. *Perancangan Dan Analisis Impeller Kompresor Diagonal Untuk Mesin Turbojet 500N.* Program Studi Aeronotika dan Astronotika Fakultas Teknik Mesin dan DIrgantara, Institut Teknologi Bandung. 2012

[7] Kutty, Hairuniza Ahmed; Rajendran, Parvathy. *3D CFD Simulation and Experimental Validation of Small APC Slow Flyer Propeller Blade*.Universiti Sains Malaysia. Penang.2017

[8] Roberts Sean C 1982 *Light Aircraft Performance For Test Pilots and Flight Test Engineers* Flight Reseach Inc, California.

IOP Conf. Series: Materials Science and Engineering **645** (2019) 012016 doi:10.1088/1757-899X/645/1/012016

# Design of Flight Control System Mechanism for National Glider GL-1

**Hendarko, T Indriyanto, M F Zulkarnain, Y K Asalani, R Duatmaja**

Faculty of Mechanical and Aerospace Engineering, Institut Teknologi Bandung
Jl.Ganesha 10, Bandung, Indonesia

Email: Hendarko@ae.itb.ac.id

**Abstract.** This paper reports a design of flight control system mechanism which will be implemented in the national glider GL-1. The flight control system in the glider GL-1 includes the following control surfaces: elevator, ailerons, and rudder for primary flight control system as well as flaps and airbrakes for secondary one. In designing the components of flight control system, selections of type, material, and dimension of components are based on loads computation as well as benchmarking to some modern gliders. Mechanism and kinematics of flight control system are designed to comply with regulation as well as space availability. Fully-mechanical flight control system is selected for the aircraft since the aerodynamic forces acting on the glider are not excessive. The aircraft uses a centre-stick as manipulator for pilot to control longitudinal and lateral movements, which is linked to elevator and ailerons by push-pull rods with coupled mechanism. Meanwhile, for directional control, an adjustable pedal is used. The pedal is linked to the rudder by tension cables. Plain flaps and flat-plat panel airbrakes are selected as flap and airbrake types in this glider. The flaps and airbrakes are controlled by mechanical levers, which are linked by push-pull rods.

## 1. Introduction

Indonesia is developing its first national glider to improve aero sport activities in the country. The glider is a single seater mid-performance sailplane with 14.3 m wing span and 7 m length of fuselage. The glider name is GL-1 and it was initially designed by a student of the Faculty of Mechanical and Aerospace Engineering ITB [1]. Since then, several studies of the glider, which includes estimation of aerodynamic characteristics and performance, structural design and analysis, as well as flight simulation, have been performed [2-6].

A half-scale prototype of the glider was built and had conducted a wing static test in 2106. In 2017 the half-scale prototype performed flight test [7]. However, the half-scale prototype had no flight control mechanism inside and remotely operated as an unmanned air vehicle. The development of the glider GL-1 is currently going into the phase of manufacturing a full-scale prototype. As a manned airplane, the full-scale prototype of GL-1 has to be equipped with flight control system. Therefore, a design of flight control mechanism which will be installed in the glider is required.

## 2. Methodology

In designing flight control mechanism of GL-1 there are some considerations that must be paid attention, i.e. space availability of the airplane, regulations referred to, and DRO of the GL-1 (especially concerning deflection range of control surfaces). In this design, the regulation referred is CS-22 [8]. Some paragraphs considered in designing the flight control system are **Subpart C - Structure** concerning **control surfaces and structures** (CS.22.395 and CS.22.397) and **Subpart D – Design and Construction** concerning **control surfaces** (CS.22.655 and CS.22.657), **control systems** (CS.22.671 to CS.22.713), and **cockpit design** (CS.22.777 and CS.22.779).

In gaining design knowledge, literature studies were accomplished in this research such as flight manual and maintenance manual of some gliders as well as some references in building home-built aircrafts [9-15]. Survey to some gliders available in Indonesia was also done to obtain a benchmark.

Flight control mechanism and kinematics are design and simulated in such a way that they comply with space availability. Whereas the selections of type, dimension, and material of the components of flight control system are based on loads computation, weight, durability, price, and availability in the market.

## 3. Flight control mechanism design

The modern high-performance sailplanes commonly use combination of push-pull rod, torque tube, and crank as the main components for control mechanism. Steel cables use is limited to rudder and tow mechanism. The design of flight control system mechanism of GL-1 includes mechanism and kinematics of elevator, ailerons, and rudder as primary flight controls as well as wing-flaps and airbrakes as secondary flight control. The design results are explained in the following subsections.

### *3.1. Control column*

A centre-stick is selected as control column for the reason of familiarity with Indonesian pilot. Movement transfer for elevator and aileron passed through the sides of the cockpit. Even though control passage through the middle of the fuselage could provide simpler design, it is not possible due to very limited space owing to the existence of tow mechanism and landing gear. Motion transfer generally uses a mechanical system, which uses a torsion bar or a crank. In the compared airplanes there is one commonly used configuration, namely the use of torsion bar to connect to the elevator mechanism. Meanwhile, to connect to the aileron mechanism uses a combination of a crank and push-pull rod. Other configurations are the use of crank and push-pull rod as in elevator mechanism on DG300 and the use of torsion bar as in aileron mechanism on ASK-21. In the design of control column for GL-1 the first option is selected because of less complexity while still fulfilling the function.

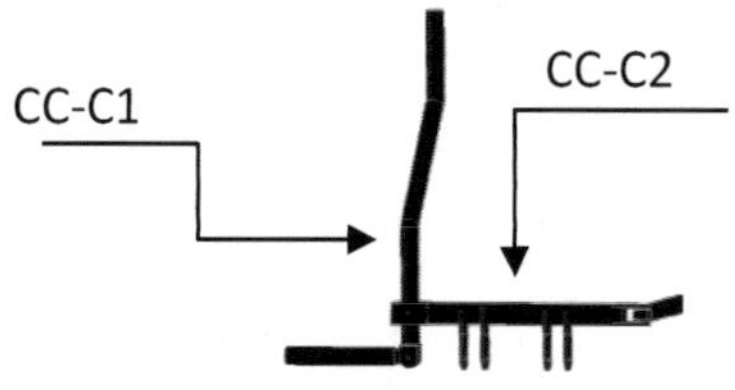


**Figure 1.** Three view drawing of control column, connected to elevator and aileron mechanism.

Control Column consists of two parts (See Figure 1).The first part of the control stick (CC-C1) works as a stick that is held and controlled by the pilot and can be moved forward- backward and left-right with a displacement of 5" or a deflection angle of the stick 25 deg. CC-C1 is connected with CC-C2 at the rotating axis, and also connected to the aileron mechanism through the lower end of the control stick. Elevator control is connected via CC-C2. The second component (CC-C2) works as a connector of the control stick (CC-C1) to control the elevator. CC-C2 is connected to the elevator rod at the right end.

### *3.2. Elevator mechanism*

The elevator uses push-pull rod mechanism. The movement transfer is designed so when the pilot pulls the stick, the elevator will be deflected upwards. A modified LAK-17 elevator mechanism [12] is selected as basic configuration for it simple mechanism. The main modification is to replace middle crank so the rod end bearings terminal that was used will be replaced with simpler and cheaper fork end terminal.

In the GL-1 elevator mechanism design, at least two crank groups are needed. One crank system is used for transferring motion from the bottom of the cockpit to the midpoint of the fuselage. Another crank system is used to change the direction from front/back movement to up/down movement at the vertical fin.

Initial sizing is done after the configuration is determined. Initial Sizing is done by estimating the location, crank dimensions, and rod dimensions based on comparison planes. The location is then fixed by considering the existence of support such as spars, torque tubes of other mechanisms, etc. The crank dimensions are then changed so that full deflection of stick gives 30 deg deflection of elevator.

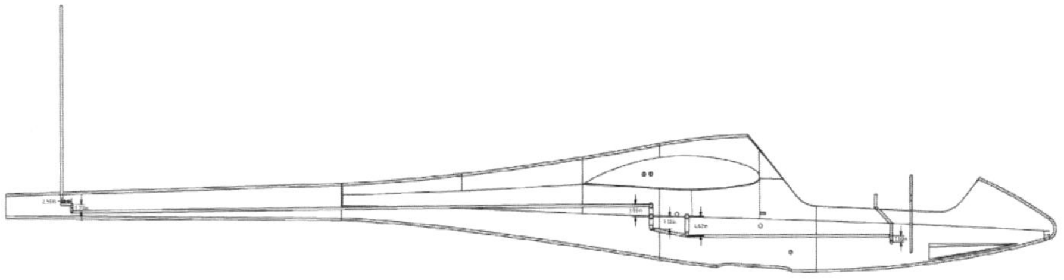

**Figure 2.** Elevator mechanism.

### *3.3. Aileron mechanism*

The aileron is a push pull-rod mechanism. It is designed so when the stick is rotated to the right, the right aileron will be deflected upwards. There are three important parts of the aileron mechanism, namely: the fuselage mechanism, the wing mechanism, and the connection.

The type of connection determines the fuselage mechanism and the wing mechanism. There are two types of connections, using torque tube or crank. In GL-1, due to the limited space in the wing, the crank type connection with vertical motion is selected. Such a mechanism is used also on Glasflügel-made planes like Libelle and Schempp-Hirth like Arcus [13]. However, the mechanism used is a mechanism specifically designed for the flaperon mechanism with airbrake flap, although it is also used in some aircrafts that have conventional spoiler configurations and aileron without flaps.

To be able to transfer the movement of horizontal translation from the control stick to a vertical translation at the connection, a rotation mechanism is required in the fuselage mechanism. To shorten the development time, the mechanism on the Libelle plane is referred with simplification accommodating the control surface configuration on GL-1. The right and left mechanisms need to have a different direction deflection, hence a mechanism that give different direction of rotation is made.

The mechanism in the wing is sufficient with a push-pull rod to the first wing segment where the aileron installed. Crank is made in such a way that the rod mechanism pulls the aileron through the underside of the wing. Initial sizing is done after the configuration is determined. Initial Sizing is done by estimating the location, crank dimensions, and rod dimensions based on comparison planes. The location is then fixed by considering the existence of support such as spars, torque tubes of other mechanisms, etc. The crank dimensions are then changed so that full deflection of stick to the right gives 10 deg up of right aileron deflection and 25 deg down of left aileron deflection, and vice versa.

In addition to the right and left mechanisms must have the opposite deflection, the aileron mechanism must also be able to produce different deflection values for up and down. This is obtained by using the crank configuration shown in Figure 3. This configuration is also used by other gliders which have similar aileron configurations. If the crank is rotated then the rotation will be transferred into vertical motion to the aileron mechanism on the wing. Because two similar cranks have different angles, one forward and one back, if both rotate in the same direction, they will have different vertical displacements. This is used to get different aileron deflection values when deflection up and down. The ratio of maximum value of up and down deflection of the aileron can be varied by changing the x angle. The results of deflection variations with a constant crank length can be seen in the table below. It should be noted that the smaller the angle, the heavier the load received by the rod. This gives a problem at the beginning of initial sizing using a 30 degree angle, the resulting load is too heavy so there is no

IOP Conf. Series: Materials Science and Engineering **645** (2019) 012016 doi:10.1088/1757-899X/645/1/012016

configuration of the chromoly tubing structure that can withstand the load. Therefore, the angle of 35 degree is selected which gives the right deflection ratio and the load value that can still be resisted by the chromoly tube structure. The deflection value is then set so that the desired deflection value is obtained.

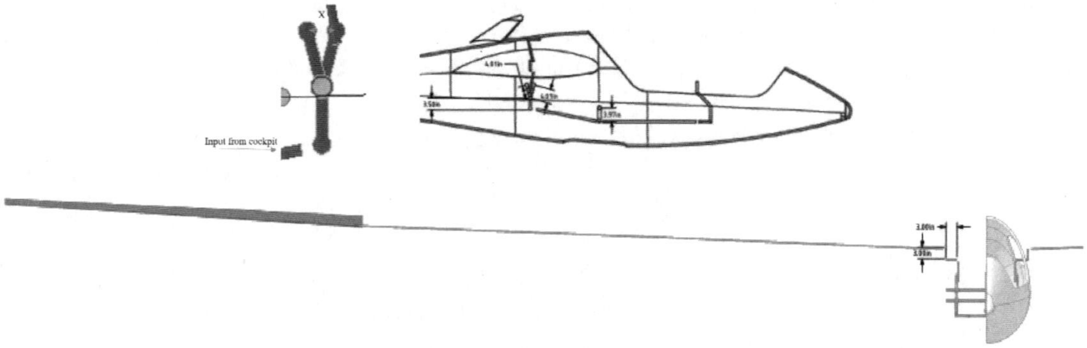


**Figure 3.** Aileron mechanism.

### 3.4. Rudder mechanism

Steel cables are commonly used for rudder mechanism of glider, which connecting pedals to the rudder. Cable mechanism is used in the design of rudder mechanism of GL-1. The need of implementing adjustable pedals becoming a reason of selecting heel-toe type pedals for this design. This type of pedal is actuated by rotating, instead of pushing, the pedal forward. It is design so that forward deflection of right pedal gives rudder deflection to the right.

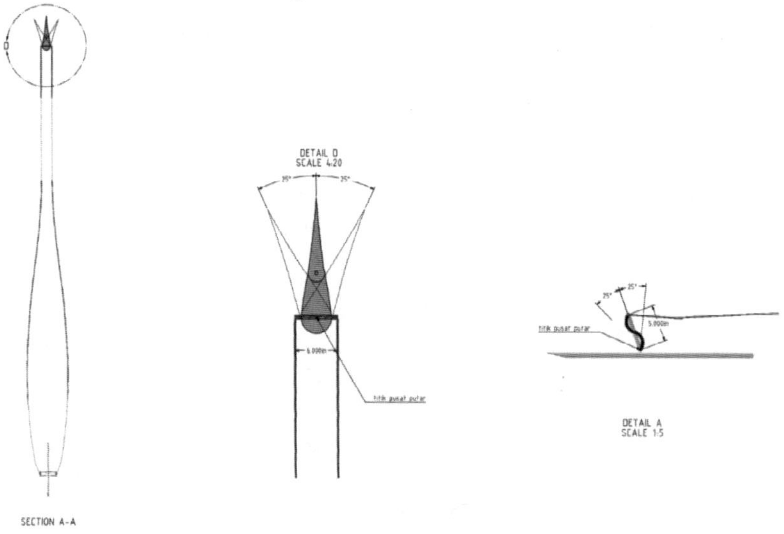


**Figure 4.** Rudder mechanism.

### 3.5. Flap mechanism

As in aileron mechanism, there are also three important parts of the flap mechanism, namely: the fuselage mechanism, the wing mechanism, and the connection.

A connection which is commonly used is a torque tube. The use of a torque tube makes it easy to synchronize the movement between the right flap and the left flap. For mechanism in the wing it is sufficient to attach the rod at the top of the torque tube to push the flap to the desired position.

The flap lever is placed at the left side of the cockpit for operation with the left hand. The pilot pulls the lever to extend the flap. This lever moves the rod and then rotates the torque tube on the flap joint.

Initial sizing is done after the configuration is determined. Initial Sizing is done by estimating the location, crank dimensions, and rod dimensions based on comparison planes. The location is then fixed by considering the existence of support such as spars, torque tubes of other mechanisms, etc. The crank dimensions are then changed so that full deflection of flap lever gives 25 deg deflection of flaps.

Initially, it is intended to have 10" of maximum lever deflection, The problem with flap mechanism is that it is difficult to reach flap lever movement to 10". The second problem is ensuring that the maximum deflection reaches 25 degrees. The main problem is that in the first crank design, there is a certain point where if it is deflected further it will cause the flap deflection to return to zero again. In this design, the point is obtained at 7.5-inch lever deflection. Alternative solution by changing the crank length is not possible because it is very close to the frame/bulkhead of the seat. Therefore it must be ascertained that the 7.5-inch lever deflection will result in 25 degrees deflection of flaps. This is also not easy because the first crank problem also appears in the third crank flap. After several iterations, the fixed geometry is finally obtained.

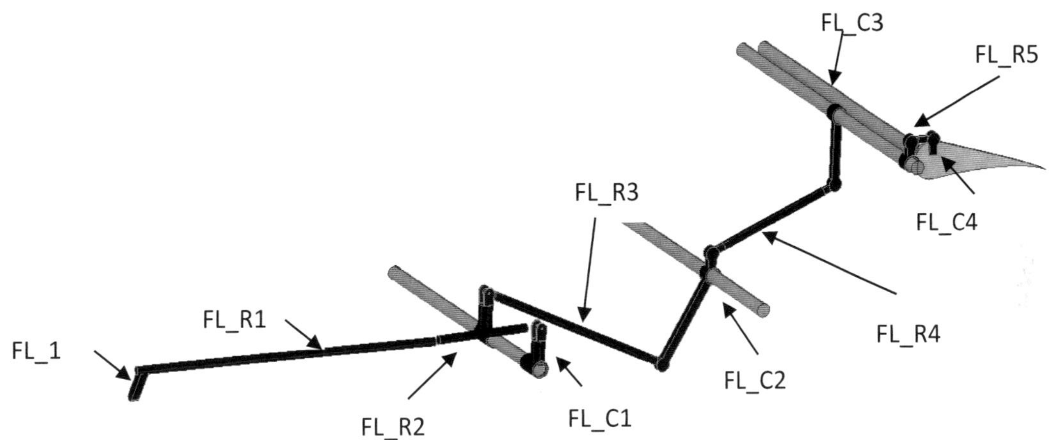


**Figure 5.** Flap mechanism.

### *3.6. Airbrake mechanism*

As in aileron mechanism, there are also three important parts of the flap mechanism, namely: the fuselage mechanism, the wing mechanism, and the connection.

Like in flap mechanism the airbrake lever is also placed at the left side of the cockpit. The pilot pulls the lever to extend the airbrake. Initially, it is intended to have 10" of maximum lever deflection. The problem with airbrake is that it is difficult to get airbrake lever deflection up to 10 inches. The main barrier is the crank shape on the fuselage. The airbrake lever deflection can be extended by extending the first arm of the first crank, but the long arm has effect of weakening structure. So far, the maximum lever deflection is 7.5 inches which brings to a full extended airbrake.

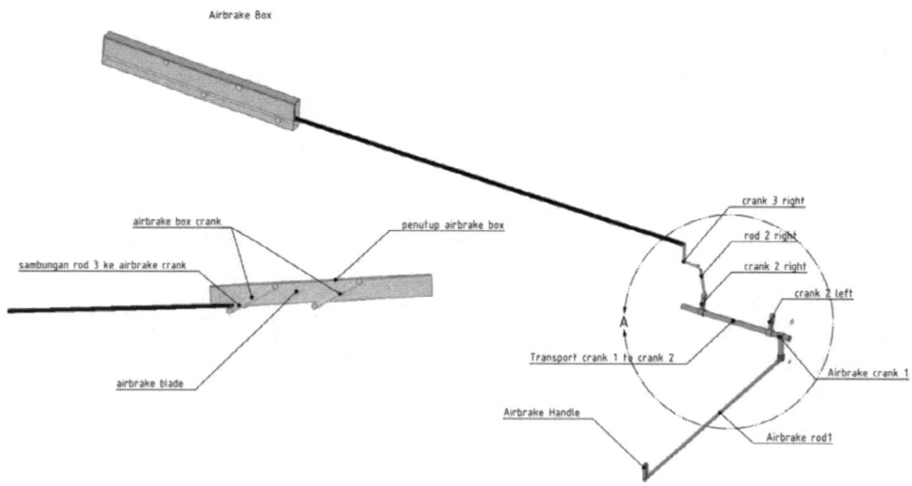


**Figure 6.** Airbrake mechanism.

## 4. Conclusion

A design of flight control system mechanism for the national glider GL-1, which includes primary and secondary controls, has been conducted. Mechanism and kinematics of flight control system are designed to comply with regulation as well as space availability. Fully-mechanical flight control system is selected for the aircraft since the aerodynamic forces acting on the glider are not excessive.

## Acknowledgement

This research was supported by P3MI ITB research scheme.

## References

[1] Hendi Pratama 2015 *Conceptual Design of Indonesia National Glider GL-01* Institut Teknologi Bandung Master Thesis
[2] Kurniasari, N. A. D. 2016 *Estimasi Parameter Aerodinamika Pesawat Glider Nasional GL-01 Melalui Reversed Engineering* Institut Teknologi Bandung Bachelor Thesis.
[3] Amalinadhi, C., Julistina, R. 2017 *GL-1 CFD Simulation Result* GL-1 Development Archives
[4] Zulkarnain, M.F. 2017 *GL-1 X-Plane Simulation Result* GL-1 Developments Archives
[5] Amalinadhi, C., Julistina, R., Moelyadi, M. A., Mulyanto, T. 2016 Comparative Study Between Schrenk and CFD Analysis for Predicting Lift Distribution Along Wing Span of Glider Aircraft. *Advances in Aerospace Science and Technology in Indonesian Vol.I*
[6] Darsono, R 2017 *Desain dan Analisis Struktur Sandwich Fuselage Pesawat Glider GL-1 Berbasis Metode Elemen Hingga dan Analitik* Institut Teknologi Bandung Bachelor Thesis
[7] Zulkarnain, M.F, Rahman, M.F., Nurhakim, M.L.I, Arifianto, O., Mulyanto, T. 2017 Flight Test of GL-1 Glider Half Scale Prototype *5th Int. Seminar of Aerospace Science and Technology*
[8] EASA 2009 *Certification Specifications for Sailplanes and Powered Sailplanes CS-22*
[9] Tony Bingelis 1992 *Sportplane Construction Techniques A Builder's Handbook*
[10] *Schweizer 1-26B Assembly Drawing*
[11] 2001 *Sailplane Maintenance Manual B1-PW-5*
[12] Sportine Aviacija 2006 *Maintenance Manual for the Self-Sustaining Powered Sailplane LAK-17AT*
[13] Schempp Hirth Flugzeugbau GmbH 2012 *Maintenance Manual for Powered Sailplane Arcus M*
[14] DG Flugzeugbau GmbH 2009 *Maintenance Manual DG-500*
[15] Rolladen-Schneider Flugzeugbau GmbH 1999 *Maintenance Manual for the LS-8a Sailplane*

IOP Conf. Series: Materials Science and Engineering **645** (2019) 012017 doi:10.1088/1757-899X/645/1/012017

# Development of small propeller test bench system

**Z S Islami[1] and F Hartono[1,*]**

[1] Faculty of Mechanical and Aerospace Engineering, Institut Teknologi Bandung, Indonesia

* Email: firman_hart@ae.itb.ac.id

**Abstract.** A set of performance analysis was performed on Master Airscrew 10X7 Electric propeller. The airfoil geometry was extracted by cutting and scanning a number of blade cross-sections. Experimentally, the test was performed at Institut Teknologi Bandung using 40 cm x 40 cm rectangular test section wind tunnel. The experimental rig was built using 3 load cells designed to measure thrust and torque. For data comparison, theoretical and numerical analysis were also done using Momentum-Blade Element method and Computational Fluid Dynamics. Compared with the previous study, the experimental data shows good agreement with most of the reference data. The theoretical data overestimated the thrust value by around 20% and underestimated the torque value by 30% at max, but the performance curve follows the similar trend with the experimental data. Both the thrust and torque value of the numerical data shows better agreement when fitted to experimental data by around 10%. While it may need more propeller samples analyzed to draw general conclusion, the three methods performed in this research showed analogous results and could be used for a particular stage of research depending on the level of accuracy needed.

## 1. Introduction

Drone usage have been significantly risen in the past few years. For small-scaled drones, propellers are popular not only for its abundance, but also its cheap price while fulfilling their function [1-2]. However, the performance charts for the smaller propellers are not usually provided, leading to trial-and-error in the propeller matching stage [3]. Brandt [1] performed a test for 79 small commercial propellers ranging from 9 to 11" while varying the RPM and wind tunnel speed to sweep over a range of advance ratios to obtain thrust and power coefficients. In this paper, an attempt was made to build a small propeller test bench that fits into wind tunnel in Institut Teknologi Bandung with a test section of 40 x 40 cm cross-section. The test bench were designed to measure thrust and torque of the propeller, while the user able to freely adjust the RPM. To ensure the results were valid, a method for calibrating the thrust and torque was developed and the test bench was used to measure a 10x7" Master Propeller Electric Propeller. The measured thrust and torque was compared with the experimental result performed by Brandt. To examine the degree of validity of theoretical and numerical method, reverse-larrabee method and Computational Fluid Dynamics model were also compared to the experimental results [4-6].

## 2. Test bench design

The test bench was built equipped by 3 main sensors to measure thrust, torque, and RPM. In Brandt's experiment, the test bench was built using an external load cell (max 10 lbs ~ 4.535 kg) to measure thrust. The torque was measured by a reaction torque transducer behind the propeller motor engine. A photoreflector was used to measure shaft RPM. Considering the configuration of existing wind tunnel in ITB and the cost needed to build the test bench, some types of sensor were picked for comparison.

IOP Conf. Series: Materials Science and Engineering **645** (2019) 012017 doi:10.1088/1757-899X/645/1/012017

### *2.1. Sensor choice*

In this research, the measurement methods take into consideration were three types for thrust, three types for torque and two types for RPM.

*2.1.1. Thrust measurement method.* To measure thrust, the three methods considered were force balance, beam deflection, and pressure difference. Using force balance method, there were two options: Internal force balance and external force balance. However, both methods have their own problems. For internal force balance, the configuration of strain gage and the complexity of development made this method costly. External force balance takes more space outside the test section it may not be suitable if the room for experiment is too cramped.

The beam deflection method utilizes beam deflection as a means of measuring thrust. Nowadays it is pretty common to use load cell strain gage to measure forces in general, usually in one direction. Due to how common the usage of load cell, it is pretty cheap and gives reasonable result within its load range provided the proper calibration.

Measuring propeller thrust using pressure difference method is measuring momentum difference in practice. The pressure behind the propeller is measured using pitot tube or five-hole probe for more accuracy. The pressure in front of the propeller is measured by the pitot tube used for wind tunnel flow velocity measurement. The thrust produced by the propeller is calculated by integrating the pressure difference with the propeller area. Although the method looks promising in theory, the flow fluctuates behind the propeller, making the pressure reading a bit more difficult without proper data logging. Considering the three options above, the load cell strain gage was chosen due to the common usage and accuracy.

*2.1.2. Torque measurement method.* The torque measurement methods considered were brake dynamometer, angle deflection and motor power measurement. Using brake dynamometer, a rotating propeller will stop accelerating rotationally given enough torque to counter it. The torque is generated by putting a brake to a disc connected to the propeller in one shaft. The force, multiplied by its radius from the shaft, will produce torque that is adjusted by the brake power. As the external force balance method, the space needed for this method is a bit much especially inside the test section. It also needs active iteration to counter the torque that makes it a bother when not automated.

If a shaft is subjected to torque, it will be twisted. The twist distance/angle may be calculated and correlated to torque with proper calibration. The reaction torque transducer used by Brandt was also made using this principle, however the torque is measured by the strain gage deflection. Knowing the torque is multiplication of force and reference length, the angle deflection could be represented by load cell deflection at a certain range.

Power is multiplication of torque and angular velocity. That means we can measure torque indirectly by measuring the motor power, that is done by multiplying the electrical current and the voltage. However, the motor itself transmit the power to the propeller imperfectly, meaning there will be some energy losses by the motor itself. If the motor efficiency along its working range is known, this method will be viable.

The torque transducer would be a better choice if not for its expensive cost. In this research, the torque generated by propeller was measured by two load cells that measure the force at a certain range from the propeller itself. The brake dynamometer would use too much space, and measuring the motor power needs the motor efficiency data.

*2.1.3. RPM measurement method.* The RPM can be measured using photoreflector or by measuring the motor frequency. Photoreflector reflects light generated by a source and gives the RPM count by the frequency it is reflected. By measuring motor frequency, 2 of 3 cable phase is used to measure the "offline" frequency, but it will need to be calibrated due to motor and propeller frequency difference. Due to compatibility with microcontroller, RPM was measured by the motor frequency.

### *2.2. Propeller test bench design result*

The test bench was made using three 3 kg capacity single-point bending load cell to measure thrust and torque. The RPM was measured by brushless RPM sensor. To connect the data to the computer, Arduino

Nano is used as Analog-Digital Converter [7]. The components used are shown in ***Table 1*** and ***Figure 1***. The design result of the test bench is shown in ***Figure 2***. For calibration, the mode with the load cell force directions and reference length is shown in ***Figure 3***. In this case, a = 3.5 cm and l = 9 cm.

**Table 1.** Components used for propeller test bench.

| No | Components | Amount |
|---|---|---|
| 1 | RCX 2836-9 880KV Outrunner Brushless Motor | 1 |
| 2 | Load Cell 3kg | 3 |
| 3 | Load Cell Driver | 3 |
| 4 | Electric Speed Control | 1 |
| 5 | Brushless RPM Sensor | 1 |
| 6 | Servo motor | 1 |
| 7 | Power Supply | 1 |
| 8 | Breadboard | 1 |

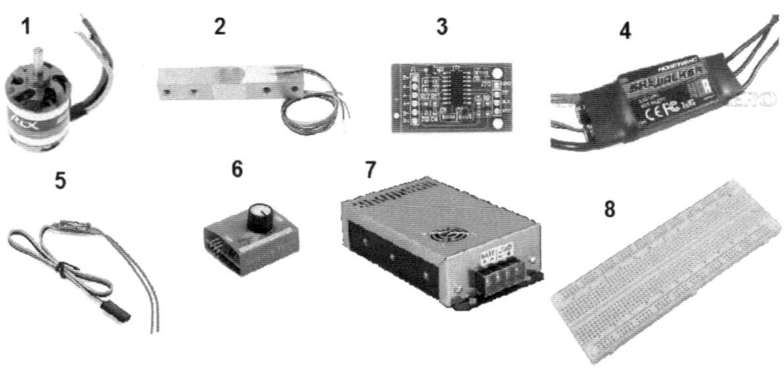


**Figure 1.** Components used for propeller test bench.

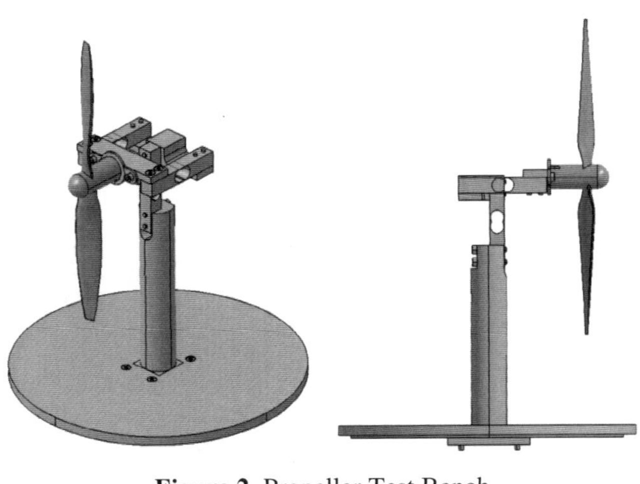

**Figure 2**. Propeller Test Bench

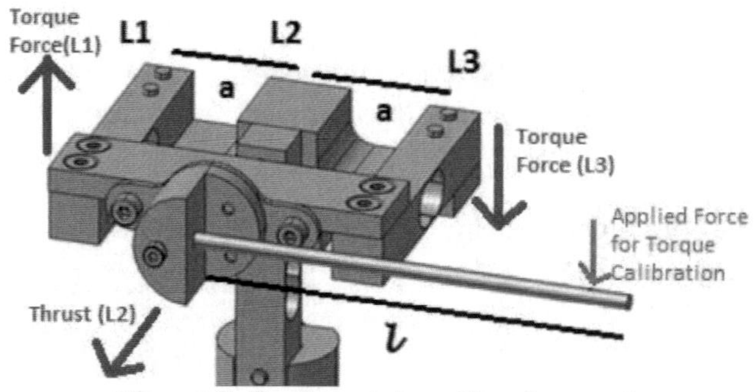


**Figure 3.** Test bench in calibration mode

## 3. Test bench calibration and results

The test bench was calibrated using applied known loads [6-8]. Originally, the load cell L2 was the only one supposed to show numbers when thrust is applied to the test bench. Likewise, the load cell L1 and L3 was supposed to be the only ones measuring the force that causes torque load. However, that is not the case as there will be some interference among the sensors or manufacturing errors. All load cells gave their readings when the test bench was subjected to either thrust or torque. Calibration was done to convert the readings into proper thrust and torque value. For thrust calibration, the loads applied are 0, 154, 254, 354, 454, 554 and 654 grams. For torque calibration, the loads applied are 0, 52, 115, 165 and 197 grams, equivalent to 0, 468, 1035, 1485 and 1773 g.cm. The calibration was done upscale and downscale 5 times.

### *3.1. Calibration results*

For thrust calibration, $L_2$ showed the most reading as intended. $L_1$ and $L_3$ gave very small reading like it was caused by interference only. For torque calibration, $L_2$ gives the smallest reading, but $L_1$ and $L_3$ showed different scale of reading although both was intended to read torque force. **Figure 4** shows the raw calibration result, and the fit from 5 times calibration is shown in **Figure 5**.

To convert raw readings to proper thrust and torque, the calibration factor matrix needed to be formed. The basic idea was to apply force and moment balance equation:

$$\sum F = 0 \tag{1}$$
$$\sum M = 0 \tag{2}$$

For thrust calibration fit, the factor used (A) is used as shown in equation (3):

$$A_1(L_1 + L_3) + A_2L_2 = T\ (gram) \tag{3}$$

The factor $A_1$ was used on both $L_1$ and $L_3$ due to their interference behavior on thrust measurement. For torque calibration fit, the factor (B) used on $L_1$ and $L_3$ were different because of the difference in calibration readings due to the difference in reference length from the applied load to the respective load cell. The length factor (K) used is defined as:

$$K = \frac{l-a}{l+a} = \frac{9-3.5}{9+3.5} = 0.44 \tag{4}$$

The basic equation used for torque is:

$$B_1(KL_1 + L_3) + B_2L_2 = Q\ (g.cm) \tag{5}$$

The calibration matrix can be defined as:

$$\begin{bmatrix} B_1K & B_2 & B_1 \\ A_1 & A_2 & A_1 \end{bmatrix} \begin{bmatrix} L_1 \\ L_2 \\ L_3 \end{bmatrix} = \begin{bmatrix} Q \\ T \end{bmatrix} \tag{6}$$

To calculate A and B, the outputs measured from the calibration fit was used. From thrust calibration fit:

$$L_1 = -0.00289481, L_2 = 0.162146671, L_3 = -0.002067894, T = 154\ gram \tag{7}$$

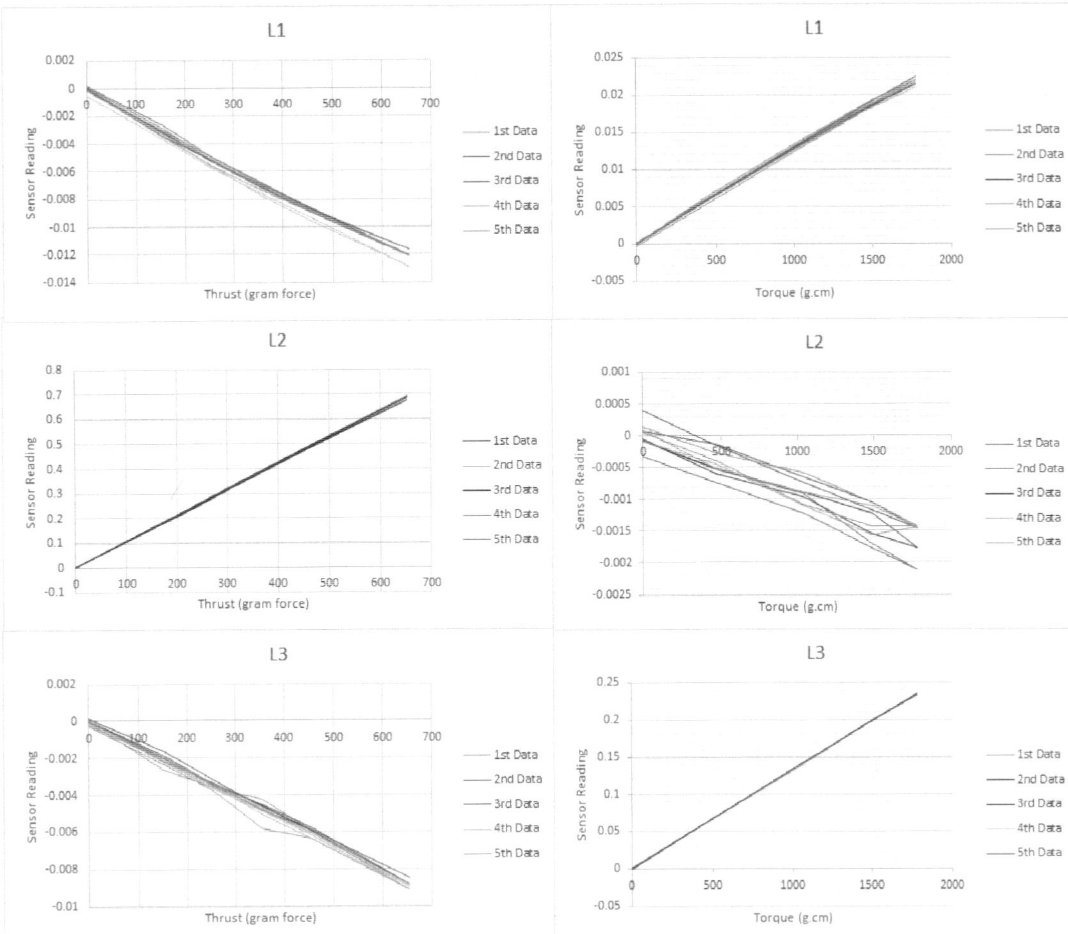


**Figure 4.** Thrust reading (left) and Torque reading (right) from calibration

$$A_1\big((-0.00289481) + (-0.002067894)\big) + A_2(0.162146671) = 154\ gram \quad (8)$$

$$B_1((0.44)(-0.00289481) + (-0.002067894)) + B_2(0.162146671) = 0\ g.cm \quad (9)$$

From torque calibration fit:

$$A_1\big((0.00578) + (0.061825)\big) + A_2(-0.00043) = 0\ gram \quad (10)$$

$$B_1((0.44)(0.00578) + (0.061825)) + B_2(-0.00043) = 468\ g.cm \quad (11)$$

Completing equation (8) and (10) yields $A_1 = 6.063$ and $A_2 = 949.943$, while completing (9) and (11) yields $B_1 = 7271.643$ and $B_2 = 149.8582$. Put into matrix (6), the calibration matrix became:

$$\begin{bmatrix} 3199.523 & 149.8582 & 7271.643 \\ 6.063 & 949.943 & 6.063 \end{bmatrix} \begin{bmatrix} L_1 \\ L_2 \\ L_3 \end{bmatrix} = \begin{bmatrix} Q \\ T \end{bmatrix} \quad (12)$$

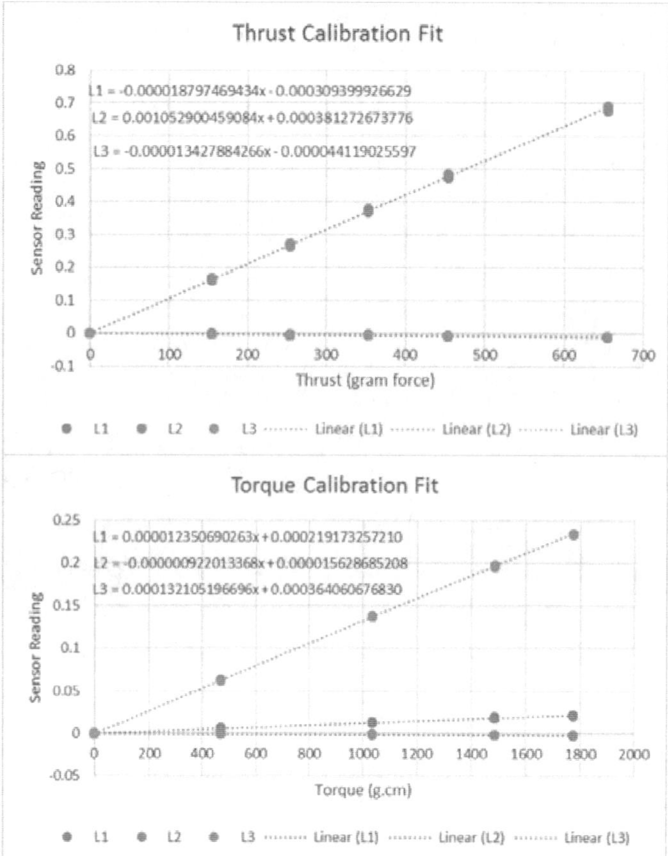


**Figure 5.** Thrust and torque calibration fit

*3.2. Experimental results compared with theoretical and numerical results*

The propeller Master Airscrew 10x7" Electric geometry was extracted to perform theoretical and numerical CFD analysis [4,7]. The theoretical analysis was done using Reverse-Larrabee method, which incorporates propeller momentum and blade element theory in nondimensional form. For the airfoil $C_l$ and $C_d$ data, the airfoil geometry from 75% radius was extracted and analyzed using CFD. For numerical CFD analysis, the model used was made to represent the experimental conditions as shown in **Figure 6**. The compared results of Brandt [1], this research experiment (Exp), theoretical (Teo) and numerical (CFD) method are shown in **Figure 7**.

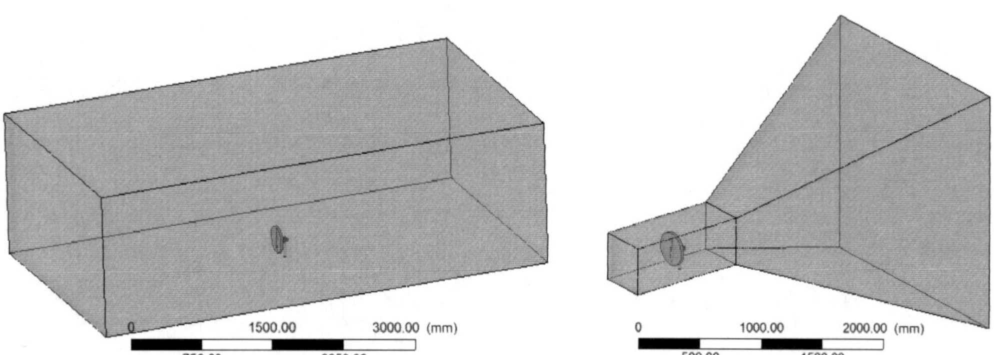


**Figure 6.** CFD model for static test (left) and tunnel test (right)

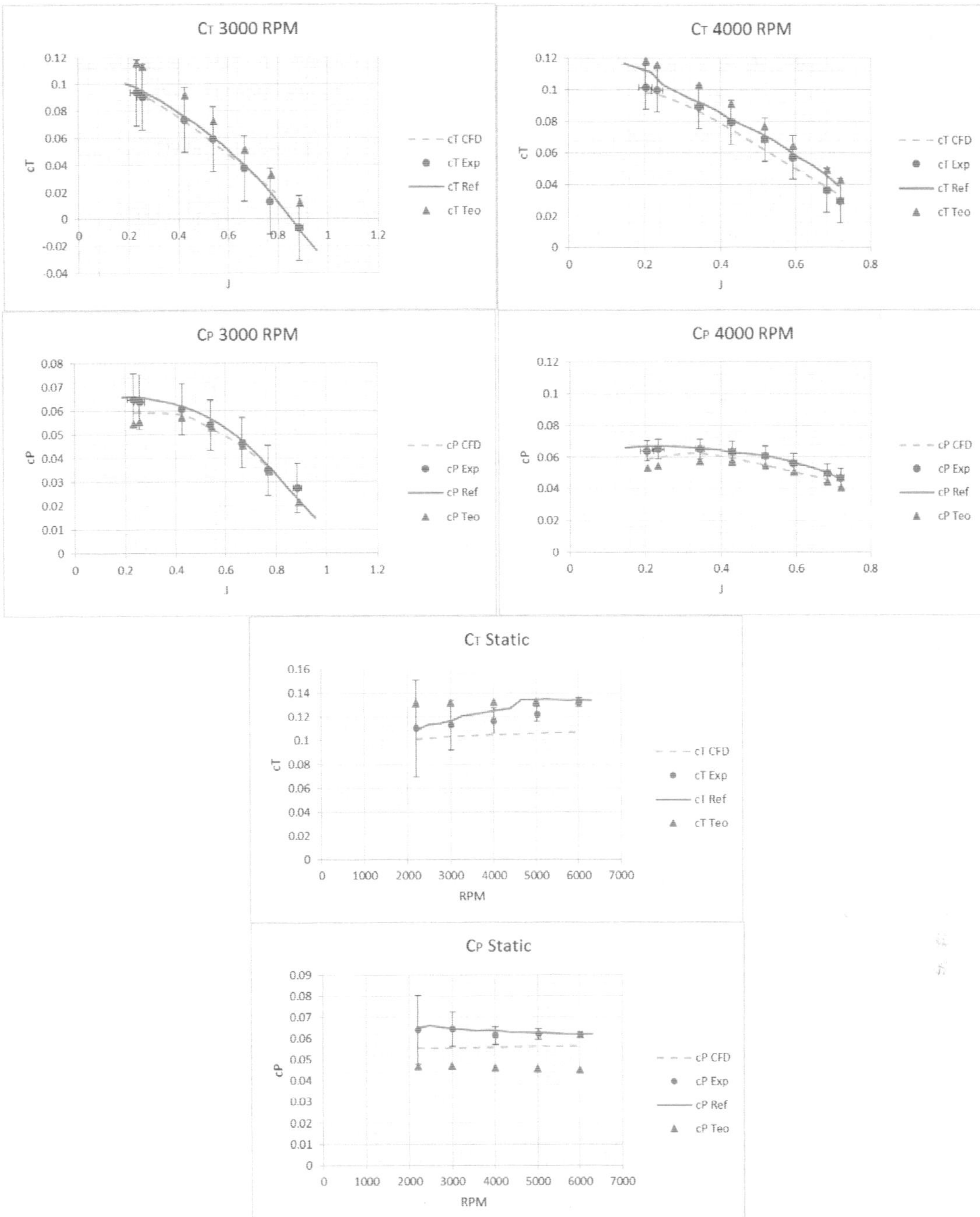


**Figure 7.** Compared results of reference, experimental, theoretical and numerical method.

The experimental results (circle) mostly shows good agreement with the reference (full-line). The CFD method (dashed line) also shows good agreement with experimental results, but the difference in results gets larger as the RPM goes up. This may be caused by the single mesh configuration used in all RPM and tunnel flow velocity, while the mesh needs to be more dense as the Reynolds number goes up. The theoretical method mostly overestimates the thrust produced and underestimates the power generated by the propeller. This may be caused by the assumption used in momentum and blade-element theory, which ignores the presence of the hub.

## 4. Conclusion

The test bench experimental results shows good agreement with the reference. This means the test bench design and the calibration method is acceptable for further use. However, the test bench is still improvable in the means of the components quality. In theory, the better the quality of the components, the more precise (the more narrow the error bars) the measurement will be, but it will be much more costly. The CFD model is also acceptable and is usable to analyze propellers of the similar case, possibly representing closely of the experimental method. The theoretical Reverse-Larrabee method seems to be falling behind in terms of accuracy. This may be caused by the degree of assumptions used. A theoretical method using a more accurate assumption may represent better in this case.

## References

[1] Brandt J B and Selig M S 2011 *Propeller Performance Data at Low Reynolds Number* (49th AIAA Aerospace Sciences Meeting)
[2] Abbot, Ira H et al 1958 *Theory of Wings Section* (New York: Dover Publication)
[3] Durand W F 1935 Aerodynamic Theory, Vol. IV (Julius Springer: Berlin)
[4] Dunn and Patrick F 2005 *Measurement and Data Analysis for Engineering and Science* (Singapore: McGraw-Hill)
[5] Carlton J S 2007 *Marine Propellers and Propulsion* (USA: Butterworth-Heinemann)
[6] Goldstein S 1929 *On the Vortex Theory of Screw Propellers* (Cambridge)
[7] Eppler R and Hepperle M 1995 *A Procedure for Propeller Design by Inverse Methods* (Stuttgart)
[8] Barlow J B Rae W H Pope A 1999 *Low-Speed Wind Tunnel Testing* (John Wiley & Sons)

IOP Conf. Series: Materials Science and Engineering **645** (2019) 012018 doi:10.1088/1757-899X/645/1/012018

# Manufacturing Fiberglass-Epoxy LSU-03 Aircraft Propeller Using Hand Lay-up and Vacuum Assisted Resin Transfer Moulding (VARTM) Methods

**A Z Dwi**[1], **H Syamsudin**[2,*]

[1]Aerospace Engineering, Faculty of Mechanical and Aerospace Engineering, Institut Teknologi Bandung

[2] Lightweight Structur , Faculty of Mechanical and Aerospace Engineering, Institut Teknologi Bandung

[*] Email: hendri.syamsudin@ae.itb.ac.id

**Abstract**. One of important aspect in designing a good quality product is on the method of how to manufacture it. In the world of aviation, products are required to meet several design criteria namely, strength, weight, and cost. Based on these design criteria, a study was carried out to manufacture LSU-03 propeller product sample using two manufacturing methods, i.e. hand lay-up and vacuum assisted resin transfer molding (VARTM). The objective is to design the manufacturing processes and to compare the quality of the propellers. The material used in this study are E-glass fiber cloth 135 and Epoxy "Lycal" resin. Prior to the propeller manufacturing process, several testing based on ASTM of manufactured composite plate was carried out to determine important parameters, i.e. thickness distribution, fiber volume fraction and voids content. A total of 3 propellers were manufactured. In order to determine the conformity of the product, propeller geometry is then measured and compared to the design mold. Finally, the comparison of the quality was made between the products of two methods. The quality of propeller products is measured based on calculated mass, thickness, void, fiber volume fraction, cost, surface quality and geometry conformity. It was concluded that VARTM is better than standard HLU and therefore more suitable for LSU-03 aircraft propeller production.

**Keywords :** Propeller, VARTM, HLU, Quality

## 1. Introduction

Many aircraft propeller manufacturers are still using machining processes for metal and wood material base. The machining process is a process of cutting the base material little by little so that the final shape is obtained according to the design. This process is expensive and long. Recently, the need for aerospace products is increasing and a low cost manufacturing method is needed to reduce costs. One suitable method is the use of fiber-polymer composites. The use of composite fibers can improve the mechanical properties whilst reduce the overall costs [1]. The process of manufacturing composite fibers is relatively simple so that mass production is possible. In addition composite materials have the advantage of better specific strength, specific stiffness and fatigue resistance [2].

In this study, the manufacturing process of propeller using Hand Lay-up (HLU) and Vacuum Assisted Resin Transfer Molding (VARTM) methods was carried out. The choice of this manufacturing method is due to its simplicity and cost competitiveness for manufacturing propeller parts.

## 2. Previous Work

Previous studies on composite manufacturing have examined variations in material properties, especially voids, fiber volume fractions, stiffness, and strength related to parameter of manufacturing process such as pressure, temperature, and other curing treatments [3-6]. Void is formed through several mechanisms. The resin flow advancing through woven or stitched fibrous preforms fills the empty regions between the fiber tows and within the fiber tows at different rates. This coupling between the two flow regimes is primarily responsible for incomplete filling of empty spaces within the fiber tows. These unfilled regions are what we term as microvoids. If the bulk flow front reaches the

IOP Conf. Series: Materials Science and Engineering **645** (2019) 012018 doi:10.1088/1757-899X/645/1/012018

vent before covering a large region of the preform, a macrovoid or dry spot is created [7]. Unlike microvoids, dry spots can be visually detected. Microvoids can also be created due to entrapment of air during resin mixing operations or due to wrinkles or pockets that are created during the lay-up [8]. Voids can be entrapped in the inter-tow spaces and are called matrix voids, or in the intra-tow spaces and are called preform voids [9]. Void is formed due to either resin cannot enter the fiber spinning due to low fiber permeability and or high resin viscosity [2,9].

### *2.1 Hand Lay-up*

HLU Manufacturing Process is the simplest manufacturing process. The resin is applied to the fiber layer until all the fiber parts are wetted. Then applying the second, third, and so on layers is carried out according to the desired number of layers. Finally, the curing process until the composite hardens. In some studies, parameter variations were carried out during the curing process. Antonio et al. [10] conducted a study of cure variations in the hand lay-up process. After the composite layer is impregnated, different treatments are carried out to achieve fully cured conditions. The results showed that curing by applying pressure resulted in a higher fiber volume fraction and a higher strength and smaller void volume. This is because by applying pressure it will push air bubbles (bubbles), which have a smaller pressure, out. Giving pressure will also suppress the composite layer so that the amount of resin will decrease so that the fiber volume fraction will rise.

In addition to the study parameters, a study of operator influence was carried out. This study was conducted in relation to the influence of the dominant operator during the impregnation process. A study by Kikuchi et al. [11] showed that the influence of the operator was unpredictable. Operators who have longer experience may not be able to perform the impregnation process faster and better than operators who do not have experience. This depends on the operator's consistency, including the direction of motion of the roll during the impregnation process, the pressure applied during the impregnation process, as well as the estimated total impregnation time by observing the gel time. With regard to strength and stiffness, the test results show that there are variations of up to 50%.

### *2.2 Vacuum Assisted Resin Transfer Moulding*

VARTM Manufacturing Process utilizes the pressure difference in the impregnation process. With the difference in pressure, the resin can flow through the fiber layer that has been suspended.

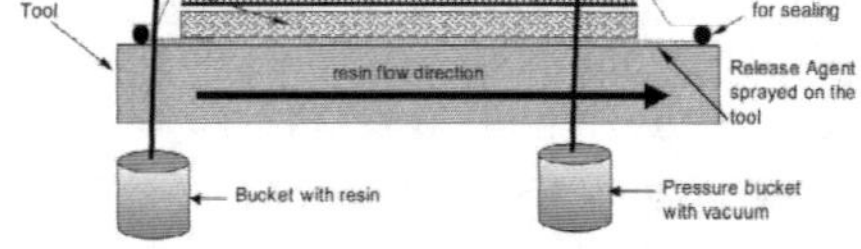


**Figure 1.** Schematic of the VARTM setup with Distribution Media [12]

In the VARTM process, vacuum condition plays a vital role. The amount of vacuum pressure given will affect the flow rate of the resin which determines how quickly and effectively the impregnation process takes place. The flow rate which is too fast, resulting in the difficulty for the resin to wet the fiber. Vishwanath et al. [5] studied the effects of pressure, vacuum and temperature with polyester / E-glass material. It is stated that the impregnation time, which is influenced by the size of the vacuum, must be adjusted with the gel time polyester, so that the curing process does not occur when impregnation is still ongoing. Yenilmez et al. [13] recommend to keep vacuum condition during curing to reduce thickness gradients due to pressure differences. Modi [14] suggested to keep the resin flowing for some time after the impregnation finished. This allows fiber yarns, which have lower permeability than fiber webbing to continue the impregnation process. However, the study mentioned that how long or how much resin must be flowed is still difficult to determine.

In addition to vacuum, there are several studies on the effect of pressure during curing in relation to layer thickness and fiber volume fraction. Pressure has a different effect when fiber is still dry, the impregnation process takes place, and the curing process. When the fiber is still dry, the pressure is limited to suppressing the fibers so that it is denser. This actually makes it difficult for the resin to flow through the fibers. During impregnation, the pressure could causes thickness variation in the composite layer, if there is a part that has not been moistened by resin or there are air bubbles that are trapped. At the curing process, the pressure will push the resin out so that the composite is densed and the fiber volume fraction rises.

## 3. Propeller Manufacturing

### *3.1 Material dan Design*

During manufacturing process, the type of fiber and resin used are the same for HLU and VARTM. Fiber type is E-glass fiber cloth 135, while the resin type is Lycal 1011. The dimensions of the propellers are following the same geometry as the original conventional wood propellers.

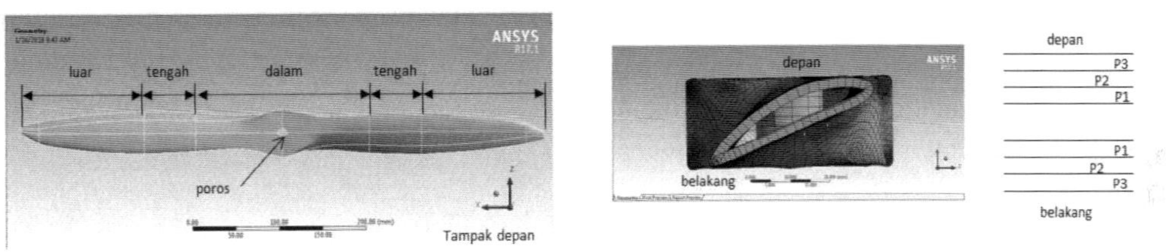


(a) (b)

**Figure 2.** Composites Propeller: (a) Design and (b) Cross section [15]

**Tabel 1.** Composites Propeller Design Configuration [15]

| Section | Orientation | Lamina Type | Number of layer | | | |
|---|---|---|---|---|---|---|
| | | | Centre | Inner | Middle | Outer |
| **P3** | ±45 | E-glass-Epoxy | 1 | 1 | 1 | 1 |
| **P2** | 0 | E-glass-Epoxy | 15 | 10 | 5 | 2 |
| **P1** | ±45 | E-glass-Epoxy | 1 | 1 | 1 | 1 |

### *3.2 Hand Lay-up Process*

The HLU process uses a brush to flatten the resin. After the resin has wetted the fiber, an emphasis is carried out to increase the fiber volume fraction.

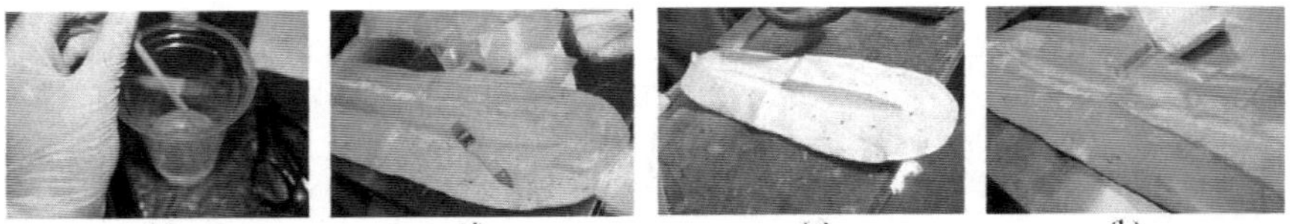

(c) (d) (a) (b)

**Figure 3.** HLU Process: (a) Preparation, (b) Lay-up, (c) Resin mixing, and (d) Impregnation

*3.3 Vacuum Assisted Resin Transfer Moulding Process*

The pressure of vacuum infusion process is maintained at 70 mm Hg by adjusting the valve of the pump. During the infusion process, the following resin conditions are maintained. Firstly, the resin stock is enough to avoid incoming air bubbles. Secondly, the resin that comes out of the outlet point is kept to be free of air bubbles. This can be done by continuing to run the pump and opening the resin inlet for a while after all the fiber parts have been moistened.

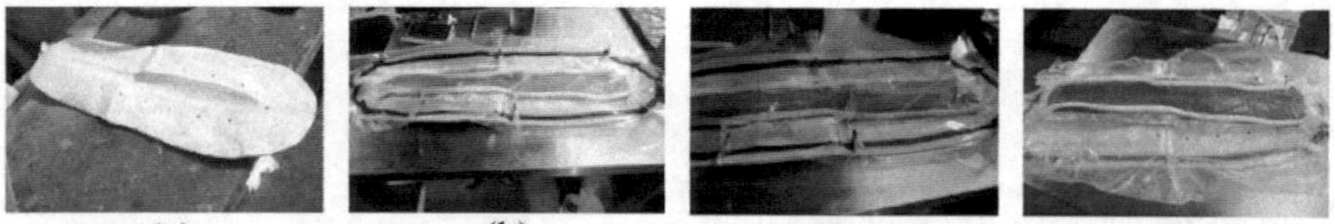


**Figure 4.** VARTM Process: (a) Preparation, (b) Vacuum Bag, (c) Infusion, and (d) De-moulding

*3.4 Assembly*

The assembly process is designed to join the upper and lower parts of the propeller. It is done by combining the upper and lower parts with the connection at the Trailling Edge (TE) and Leading Edge (LE), using cold bonding technique. The material used as the join is a woven E-glass placed at the orientation of ± 45.

**Figure 5.** Joining design at TE and LE

## 4. Analysis of propeller quality

*4.1 Mass comparison*

Using Krisbow KW0600377 Precision Scale, the results show that the propellers manufactured by the VARTM method are 24% lighter than HLU.

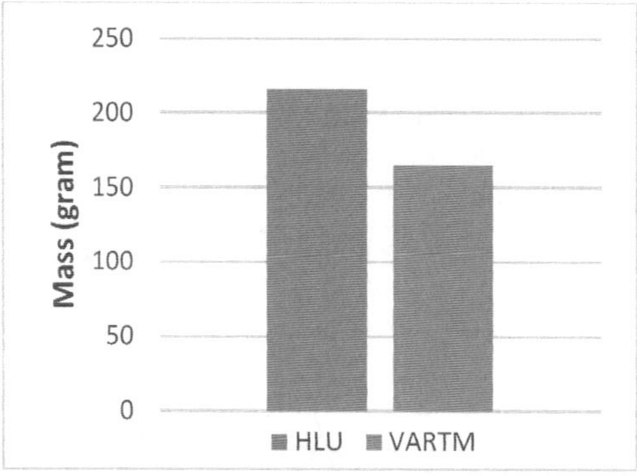


**Figure 6.** Mass measurentment result

Mass measurement results are in line with the amount of pressure given during the curing process. At VARTM the pressure given is about 2 atm while HLU is 1 atm. With higher pressure, the amount of resin will decrease and automatically the total mass will decrease.

IOP Conf. Series: Materials Science and Engineering **645** (2019) 012018 doi:10.1088/1757-899X/645/1/012018

### *4.2 Fibre Volume Fraction and Void Comparison*

Good mechanical properties of composite structure can generally be known from low percentage of voids and high percentage of fiber volume fractions [4]. Measurement of void and fiber volume fraction refers to ASTM 2734-09 with the void measurement referring to ASTM 3171-99.

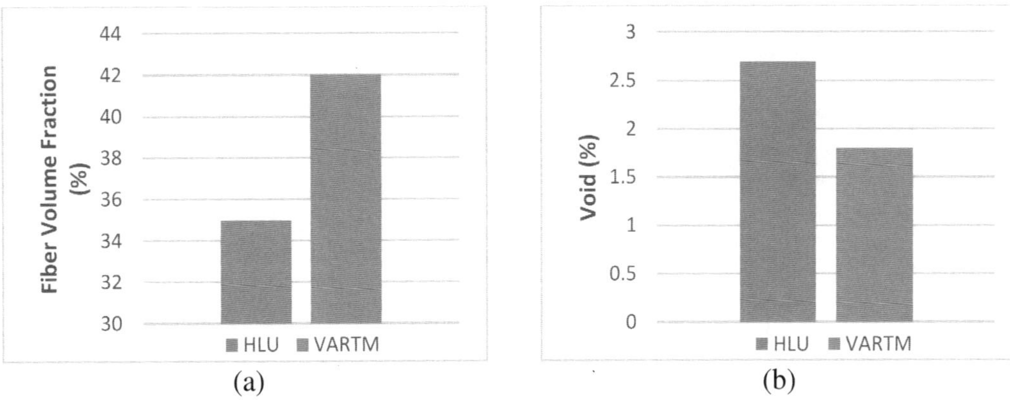


**Figure 7.** Result of: (a) Fiber Volume Fraction and (b) Void

Measurement of fiber volume fraction shows that VARTM has a fraction of 20% higher. While VARTM voids are 34% lower.

### *4.3 Geometry Conformity*

Conformity to design geometry at the various position of the propeller is shown in the following figure. It can be seen that the two manufacturing methods achieve around 90% of design geometry.

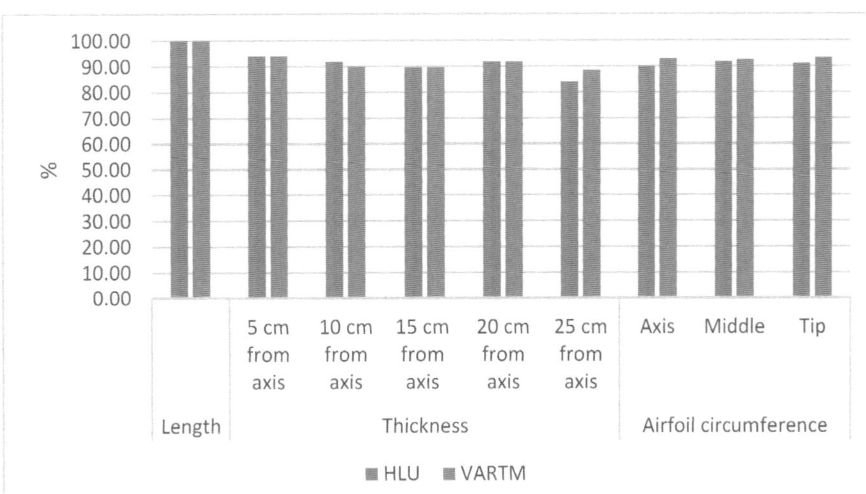


**Figure 8.** Product convormity

### *4.4 Cost Comparison*

Costs can be divided into investment costs and production costs. Investment costs are costs for equipment that can be used for a period of time, such as pump, jig, fixture etc . While production costs are the costs needed for production, such as consumable material, fibre, resin, salary, electricity, etc. During this study, for manufacturing three propellers, the cost of using VARTM method is double of the HLU. However, it was found that the cost of production using VARTM is reduced with higher number of propellers.

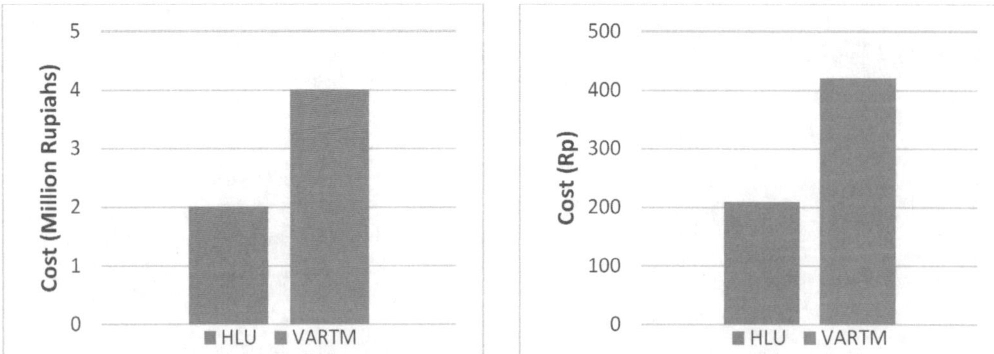


(a) (b)

**Figure 9.** Calculation of: (a) Initial Investment and (b) Production cost

*4.5 Assembly*

Generally good, there are only a few miss geometries in the LE and TE sections because the number of layers is too thick. With this condition, it is certain that there will be a miss aerodynamic calculation. One way to overcome this is by sanding.

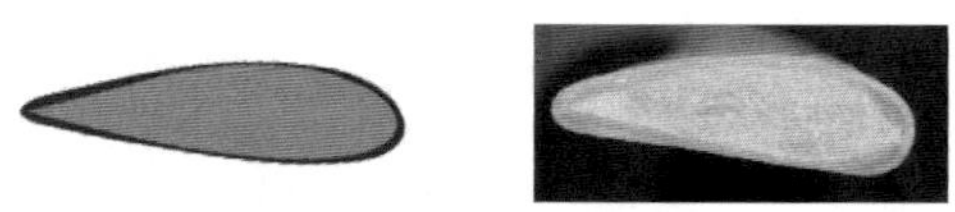

(a) (b)

**Figure 10.** Assembly: (a) Prediction and (b) Result

*4.6 Balancing*

A good propeller must have a center of mass right on the shaft (center). This is to avoid mass imbalances which result in vibration when the propeller rotates. Continuous vibration can cause the loss of propeller power to cause the propeller to break. Also to maintain the balance of the lift force associated with symmetry between the sides of the propeller. In the LSU-03 propeller manufacturing, the product that has been completed at the assembly is carried out in a simple equilibrium analysis [15]. By using an iron plate to get the center of mass. And the results show that the center point of the propeller mass is around the shaft. This indicates that the propeller is balanced and can be used for further testing purposes (rotary / vibration test, aerodynamic test, and structural test).

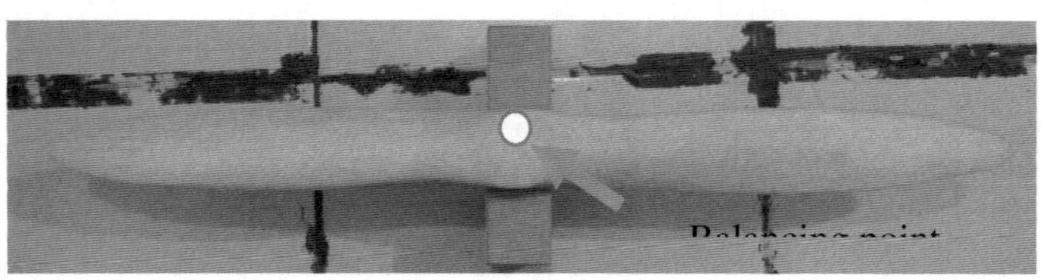

**Figure 11.** Balancing point

## 5. Discussions

In the aerospace world, two most important aspects are quality and cost. The quality of VARTM products is superior. This can be seen from a lighter mass of around 24%, a lower void of about 34%, and a higher fiber volume fraction of around 20%. This is the main point in the assessment of composite products. VARTM products are lighter due to greater pressure during curing than the HLU process. Thus the amount of resin contained in the final product will be less. With the same amount of fiber, the VARTM product produces lighter mass. While voids are related to the presence of vacuum during the curing process. In the VARTM process, the pressure is maintained at a condition of around 0 atm so that the possibility of air being trapped in the composite layer is smaller. Which also reduces the possibility of voids. The higher fiber volume fraction in VARTM products is again due to the higher pressure on the curing process so that it reduces the amount of resin. With a fixed amount of fiber, the fiber volume ratio to the total composite volume will be higher.

In terms of production, namely costs, VARTM requires approximately 2 times the cost of HLU both for awall investment and the cost of producing datu. This is quite reasonable because VARTM method uses several materials that cannot be used repeatedly such as bagging films, breather, and peel ply if production is only done once. With mass production, there is a possibility that the VARTM method produces products with a cost that may be only 20% more expensive than HLU products. While in terms of geometry, both HLU and VARTM have a geometry match of about 90-95%. This indicates that the level of trust, especially the aerodynamic aspects of the propeller reaches 90-95%. In terms of mass balance, both VARTM and HLU have produced good products.

To determine the method suitable for the LSU-03 propeller manufacturing process, quality parameters (mass, mechanical properties, and geometry), production (cost), and ease of manufacturing processes are used [16].

**Tabel 2.** Comparasion result

| Parameter | Max | HLU | VARTM |
|---|---|---|---|
| **Thickness** | 10 | 8 | 10 |
| **Mass** | 15 | 10 | 15 |
| **Strength** | 15 | 10 | 15 |
| **Repeatibility** | 9 | 7 | 9 |
| **Surface quality** | 10 | 8 | 8 |
| **Production Time** | 10 | 10 | 9 |
| **Investment** | 8 | 8 | 5 |
| **Production Cost** | 15 | 15 | 10 |
| **Accesability** | 8 | 8 | 5 |
| **Total** | **100** | **84** | **86** |

## 6. Conclusions

In the composite manufacturing process, there are parameters that affect both the manufacturing process and the final product. In general these parameters are pressure and vacuum. Pressure has a role to drive the resin so that the fiber volume fraction rises. While vacuuming is related to the number of voids. The lower the vacuum, the lower the voids. In this study, the manufacturing process of fiber cloth 135 E-glass / Epoxy using material HLU and VARTM was carried out. The study results show that VARTM products are superior in terms of quality while HLU is superior in terms of costs and manufacturing processes. The suitability of HLU and VARTM products is quite good, in the range of 90-95%. The center point of both HLU and VARTM propeller mass is good. By conducting comparative and assessment studies covering the quality, cost, and other aspects of production with the point-point assessment method, it can be concluded that the VARTM method is a more suitable method for manufacturing aircraft propellers LSU-03.

**Acknowledgements**
The authors would like to acknowledge the financial support from Ministry of Research, Technologyand Higher Education (Kemenristekdikti) Indonesia under the scheme of Penelitian Terapan Unggulan Perguruan Tinggi, 2018.

**References**

[1] Shama Rao N., Simha T.G.A., Rao K.P. and Ravi Kumar G.V.V. 2018 Carbon Composites are Becoming Competitive and Cost Effective. *Infosys.*
[2] Susanna Laurenzi and Mario Marchetti 2012 Advanced Composite Materials by Resin Transfer Molding for Aerospace Applications. *INTECH.* 197-226.
[3] F.Y.C. Boey and S.W. Lye 1990 Effect of Vacuum and Pressure in an Autoclave Curing Process for a Thermosetting Fibre-Reinforced Composite. *Journal of Materials Processing Technology.* **23** 121-131.
[4] Nina Kuentzer, Pavel Simacek, SureshG. Advani, and Shawn Walsh 2007 Correlation of Void Distribution to VARTM Manufacturing Technique. *Composites: Part A.* **38** 802-813.
[5] Viswanath R. Kedari, Basil I. Farah and Kuang-Ting Hsiao 2011 Effects of Vacuum Pressure, Inlet Pressure, and Mold Temperature on theVvoid Content, Volume Fraction of Polyester/ E-glass Fiber Composites Manufactured with VARTM Process. *Journal of Composite Materials.***45** 2727-2742.
[6] Kundavarapu Vengalraoa, Kopparthi Phaneendra Kumarb, Dasari Venkata Ravi Shankerc, Nadendla Srinivasababud and Aerra Kiran Kumar Yadav 2016 An Investigation on the Quality of the Laminates Produced by VARTM Process and Process parameters. *Materials Today: Proceedings.* **4** 9196-9202.
[7] Liu B, Bickerton S, Advani SG 1996 Modelling and simulation of resin transfer moulding (RTM) – gate control, venting and dry spot prediction. *Compos Part A: Appl Sci Manuf* **27(2)**135–41.
[8] Avila AF, Morais DTS 2005 A multiscale investigation based on variance analysis for hand lay-up composite manufacturing. *Compos Sci Technol* **65(6)** 827.
[9] Hamidi YK, Aktas L, Altan MC 2005 Three-dimensional features of void morphology in resin transfer molded composites. *Compos Sci Technol* **65(7–8)** 1306.
[10] Antonio F. Avila and David T.S. Morais 2005 A Multiscale Investigation Based on Variance Analysis for Hand Lay-up Composite Manufacturing. *Composites Science and Technology.* **65** 825-838.
[11] T. Kikuchi, H. Hamada, A. Naki, A. Ohtani, A. Goto, Y. Takai, A. Endo, C. Narita, T. Koshino, and A. Fudauchi 2013 Process Analysis of Hand Lay-up Method by Various Experience Persons. *The 19$^{TH}$ International Conference on Composite Materials.*
[12] Dominik Bender, Jens Schuster, and Dirk Heider 2006 Flow Rate Control During Vacuum-assisted Resin Transfer Molding (VARTM) Processing. *Composites Science and Technology.* 66 2265-2271.
[13] Yenilmez B, Senan M and Sozer EM 2009 Variation of part thickness and compaction pressure in vacuum infusion process *Compos Sci Technol* **69** 1710–1719.
[14] Modi D. 2003 *Numerical issues in mold filling simulations for composites processing*. (Newark: University of Delaware)
[15] Awang Nuranto 2018 *Simulasi & Analisis Desain Struktur Baling Baling Pesawat LSU-03 menggunakan Material Komposit* (Bandung: Institut Teknologi Bandung).
[16] Sanjay K. Mazumdar 2002 *Composites Manufacturing : Materials, Product, and Process Engineering* (Florida: CRC Press)

IOP Conf. Series: Materials Science and Engineering **645** (2019) 012019 doi:10.1088/1757-899X/645/1/012019

# Attitude and flight path control systems of unmanned combat aerial vehicle

**A Hanif[1,*] and R A Sasongko[1]**

[1] Institut Teknologi Bandung, Faculty of Mechanical and Aerospace Engineering, Jl. Ganeca 10, Bandung, Indonesia

[*] E-mail : azizulaah@s.itb.ac.id

**Abstract.** UCAV as one of the UAV application needs to have good maneuverability to be able to carry out missions. For this reason, an Automatic Flight Control System (AFCS) is needed to facilitate the operator in controlling the maneuvering motion. This study aims to design an automatic flight control system that is able to track the given maneuver control commands so that the aircraft can maneuver according to the control command. First, an analysis of the open loop system is conducted to determine the characteristics of the UCAV to be controlled. The stability augmentation system is then designed to improve the dynamic stability characteristics of UCAV. Finally, a command augmentation system is created that uses the tracking controller principle. This principle aims to track control commands (attitude and flight path variables to control maneuvers) so that the aircraft is able to maneuver according to the given control command. The simulation results show that the automatic flight control system that is being made has been successful in tracking the maneuver control command with the optimal method.

## 1. Introduction

Unmanned Aircraft Vehicle (UAV) - are airplanes that are operated without the presence of human pilots in the aircraft (on-board) but controlled by operators from land or an automated system. Like aircraft, UAV can also be used for military purposes. UAV that is used for military purposes are called Unmanned Combat Aerial Vehicle (UCAV). UCAV is commonly used for surveillance, ground attacks using either missiles or bombs. The mission profile of UCAV is shown in Figure 1.

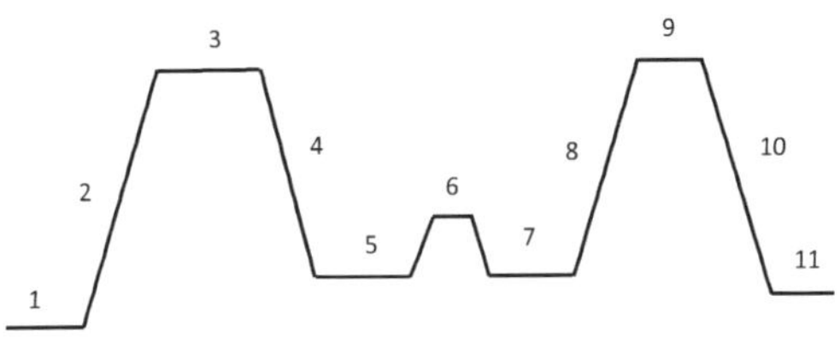


| | Mission type | | Mission type |
|---|---|---|---|
| 1 | Take-off | 7 | Dash out |
| 2 | Climb/Accelerate | 8 | Climb/Decelerate |
| 3 | Cruise | 9 | Cruise |
| 4 | Descent/Accelerate | 10 | Descent/Decelerate |
| 5 | Dash in | 11 | Landing |
| 6 | Combat (weapon drop) | | |

**Figure 1.** UCAV Mission Profile. [1]

Therefore, an automatic system is needed to control the vehicle when maneuvering. This automated system is also called an Automatic Flight Control System (AFCS). One of the roles of AFCS is to track the given control commands (in this case the maneuvering motion) so that the aircraft performs the motion according to the control commands given. In this paper control command given only in longitudinal mode. Variables which are used as maneuver commands are flight path angle ($\gamma$) and pitch angle ($\theta$). Control

IOP Conf. Series: Materials Science and Engineering **645** (2019) 012019 doi:10.1088/1757-899X/645/1/012019

commands that are being given are zero flight path angle while pitch angle varies along the simulation. This command is represented pitch pointing maneuver that useful for directing missile or another system to desired direction without changing flight path and another variables significantly.

The purpose of this paper is to construct a control system i.e. command augmentation system so that the aircraft can track the given control commands and be able to perform certain maneuvers but still notice the limit of UCAV so that the response generated from the simulation can be performed by UCAV in real circumstance.

## 2. Control system structure

This section discusses the structure of control systems which are being used in this paper. The structure uses observer system and tracking controller principle. For the inner loop, stability augmentation system (SAS) is used to increase aircraft dynamic characteristic. SAS is designed by using state feedback with observer because the output variable isn't as same as state variable. Then, outer loop is designed to track the control command with command augmentation system (CAS).

The block diagram which contains observer and tracking controller is shown in Figure 2.

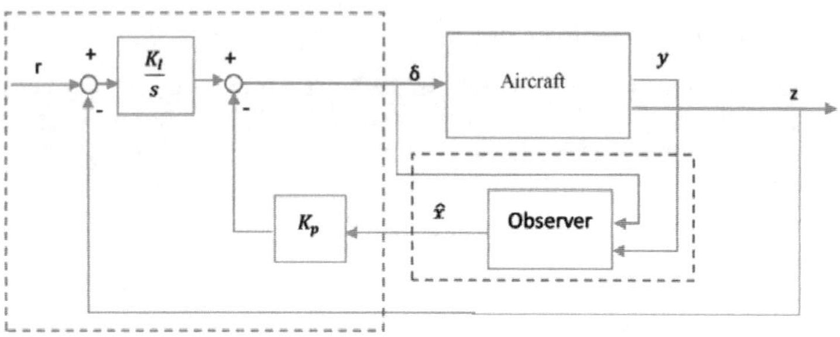


**Figure 2.** Control system structure with tracking controller and observer

Red square border indicates tracking controller while purple square border indicates observer system.

In this case with γ-θ control, SAS in longitudinal is used for inner loop with y = q, $\bar{x} = [u\ w\ q\ \theta]^T$, control signal $\delta_E$ and $\delta_T$, $z = [\ \gamma\ \theta\ ]^T$, and $r = [\ \gamma_{ref}\ \theta_{ref}\ ]^T$.

### 2.1. Tracking controller

Reference signal that being used on the aircraft maneuver control can have varying values depend on the desired maneuver command. A system which allows the variable on aircraft following the reference command that vary with time is called tracking controller [2].

If a new variable $\dot{e}$ is defined as

$$\dot{\boldsymbol{e}} = \boldsymbol{z} - \boldsymbol{r} \quad (1)$$

where z is output response and r is reference signal so that new state variable is formed by adding error state as

$$\boldsymbol{X_c} = [\boldsymbol{x}\ \boldsymbol{e}]^{\boldsymbol{T}} \quad (2)$$

Then, control input signal equation become

$$\boldsymbol{u} = \widetilde{\boldsymbol{K}}\boldsymbol{X_c} \quad (3)$$

$\widetilde{K}$ variable is separated become

$$\widetilde{\boldsymbol{K}} = [\ \boldsymbol{K_p}\ \boldsymbol{K_I}] \quad (4)$$

where $K_p$ is given by the first n columns of $\widetilde{K}$ with the same size as matrix A and $K_I$ is given by the last columns according to the number of state error.

By doing substitution equation (2) and (4) to equation (3) the new equation is obtained as

$$u = K_p x + K_I e \tag{5}$$

Equation (5) can be rewrited as

$$u = K_p x + K_I \int z - r \tag{6}$$

Equation (6) is generalization of classical PI (Proportional Integral) controller.

As a result, a new state space formulation is obtained that called augmented state space with the equation as

$$\dot{\bar{x}}_c = A_{aug}\bar{x}_c + B_{aug}\bar{u} \tag{7}$$

$$\mathrm{A_{aug}} = \begin{bmatrix} \mathrm{A} & 0 \\ \mathrm{C} & 0 \end{bmatrix}; \mathrm{B_{aug}} = \begin{bmatrix} \mathrm{B} \\ 0 \end{bmatrix}$$

$$z = C_{aug}\bar{x}_c \tag{8}$$

$$\mathrm{C_{aug}} = [C \quad 0]$$

where z is a variable which desired to be controlled so the response is tracking the reference.

### 2.2. *Observer/estimator*

In general, it is assumed that all the states can be sensored for feedback. But in fact, not all the state can be sensored because the availability and the accuracy level of the sensor. To overcome the problem,there must be a system that can estimate state variables that can be sensored. This system is called observer. Observer utilizes the information from input and the output system to obtain estimated variable $\hat{x}$ that fed to feedback controller. Gain for observer is acquired with pole placement technique so that the estimated variable is obtained. In longitudinal mode, $\hat{x} = [\hat{u}\ \hat{w}\ \hat{q}\ \hat{\theta}]^T$.

## 3. Control method

The following control methods are presented to control the linear model of UCAV.

### 3.1. *Pole Placement*

The pole placement technique begins with the determination of the desired closed loop poles based on the frequency response requirement such as the damping ratio. By determining the desired closed loop poles at $s = \mu_1$, $s = \mu_2$, …, $s = \mu_n$ .By selecting an appropriate gain matrix for state feedback, it is possible to force the system to have a closed loop poles at the desired location if the original system is fully controllable[3]. Pole placement gain is obtained by using MATLAB command 'K= place (A,B,p)'.

### 3.2. *Optimal control using LQR*

Linear quadratic regulator (LQR) is one of the optimal control techniques, which takes into account the states of the dynamical system and control input to make the optimal control decisions. Gain K that being used is derived from minimization of the cost function [4].

$$J = \int_0^\infty (x^T Q x + u^T R u)\, dt \tag{9}$$

where Q and R are positive semi-definite and positive definite symmetric constant matrices, respectively. The LQR gain vector K is given by

$$K = R^{-1} B^T P \tag{10}$$

where is a positive definite symmetric constant matrix obtained from the solution of matrix algebraic Riccati equation (ARE).

IOP Conf. Series: Materials Science and Engineering **645** (2019) 012019 doi:10.1088/1757-899X/645/1/012019

## 4. Dynamic model of UCAV

In this section, the dynamic model of UCAV is presented. First, the specification of the UCAV is shown in Table 1.

**Table 1.** UCAV Specification

| General characteristics | | Performance | |
|---|---|---|---|
| Length | 11 m | Maximum speed | 260 kts (482 km/h) |
| Wingspan | 20 m | Cruise speed | 169 kts (313 km/h) |
| MTOW | 4760 kg | Range | 1000 nm (1852 km) |
| Fuel capacity | 1800 kg | Endurance | 14 hours fully loaded |
| Payload | 1700 kg | Service ceiling | 50000 ft (15000 m) |
| Powerplant | 1 Honeywell TPE331-10 turboprop 900 hp(671kW) | Operational altitude | 25000 ft (7500 m) |

From the specification on Table 1, dynamic model of UCAV is generated in state space form. This model is obtained from reference [3].

Dynamic model of UCAV (h = 25000 feet (7600m) $V_{trim} = 87$ m/s)

$\bar{x} = [u \quad w \quad q \quad \theta]^T \qquad \bar{u} = [\delta_e \; \delta_T]^T$

$$\boldsymbol{A} = \begin{bmatrix} -0.05 & 0.17 & -11.1 & -9.7 \\ -0.1 & -0.75 & 85.2 & -1.3 \\ 0.003 & -0.04 & -0.24 & 0 \\ 0 & 0 & 1 & 0 \end{bmatrix} \quad \boldsymbol{B} = \begin{bmatrix} 0.72 & 0.002 \\ -5.4 & 0 \\ -24.2 & 0 \\ 0 & 0 \end{bmatrix} \tag{11}$$

Then the dynamic characteristic of UCAV model is presented in Table 2.

**Table 2.** UCAV Dynamic Characteristic

| Poles location | Damping ratio |
|---|---|
| -0.5 ± 1.84 | 0.261 |
| -0.0014 ± 0.13 | 0.0108 |

Based on Table 2 it can be seen that the dynamic characteristic of UCAV model is needed to be improved.

## 5. Control system design

In Table 3 it can be seen that the UCAV model's dynamic characteristic need to be improved or stabilized. So, this paper uses Stability Augmentation System (SAS) which has function to improve the stability of aircraft by increasing the damping ratio. This paper also uses SAS by giving state feedback on the system.

CAS then designed to make the aircraft response same with the control command given. As previously explained, CAS has two variables which will be given feedback by the tracking controller, namely γ angle and θ angle. So, there will be two new state variables in the form of an error variable with sequence that the first error variable is the variable γ and the second is the variable θ. The formulation for this case as described in subsection 2.1 with variables $X_c = [u \; w \; q \; \theta \; e_1 \; e_2]^T$, $y = z = [\, \gamma \; \theta \,]^T$.CAS is designed by LQR. In design controller process, Q and R values are varied to get better results. Gain in CAS is obtained by Q configuration and R as follows.

$$Q = \begin{bmatrix} 1 & 0 & 0 & 0 & 0 & 0 \\ 0 & 1.2 & 0 & 0 & 0 & 0 \\ 0 & 0 & 1 & 0 & 0 & 0 \\ 0 & 0 & 0 & 1.2 & 0 & 0 \\ 0 & 0 & 0 & 0 & 5 & 0 \\ 0 & 0 & 0 & 0 & 0 & 750 \end{bmatrix} \qquad R = \begin{bmatrix} 10 & 0 \\ 0 & 10 \end{bmatrix}$$

Gain for CAS is shown in Table 3.

**Table 3**. Gain of CAS

| K1 | K2 | K3 | K4 | K5 | K6 |
|---|---|---|---|---|---|
| 0.04 | -0.14 | -1.56 | -17.67 | 0.005 | -8.66 |
| 0.08 | 0.007 | 0.0008 | 0.29 | 0.71 | 0.06 |

K1 to K4 values are used as feedback gain on SAS, while K5 and K6 are used as tracking controllers gain.

## 6. Simulation and result

Before showing the simulation result, first it will be shown the control command given. Control commands given in longitudinal mode are pitch angle with variation value which is shown in figure while flight path angle is is held constant by zero.

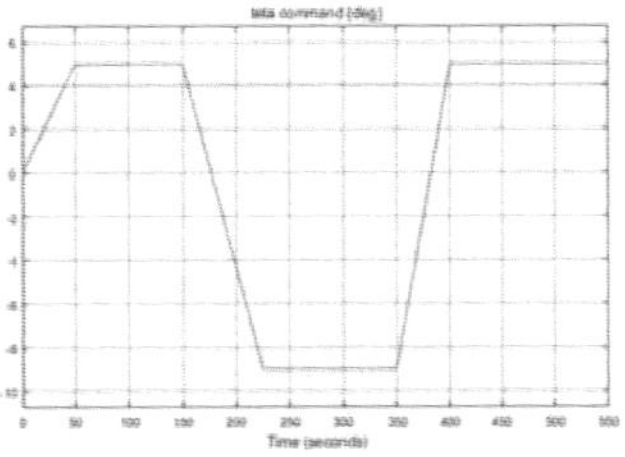


**Figure 3.** Pitch angle command

Illustration of the aircraft movement as a result of control command is presented in Figure

**Figure 4.** Aircraft movement illustration

By using gain in Table 3, aircraft tracking response, output response, and control surface response of CAS are shown in Figure 5 and Figure 6.

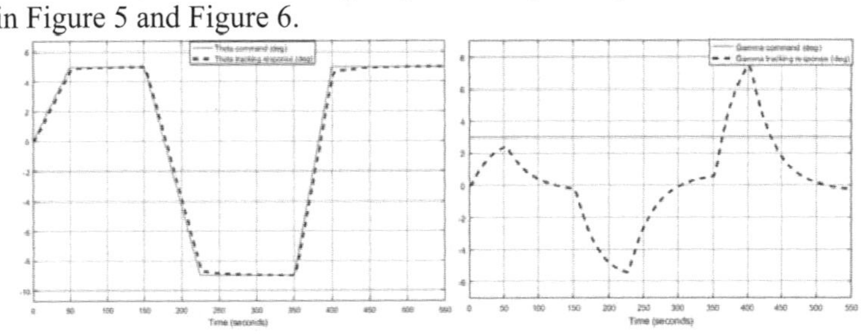


(a) (b)

**Figure 5.** Result of CAS tracking response (a) pitch angle (b) flight path angle

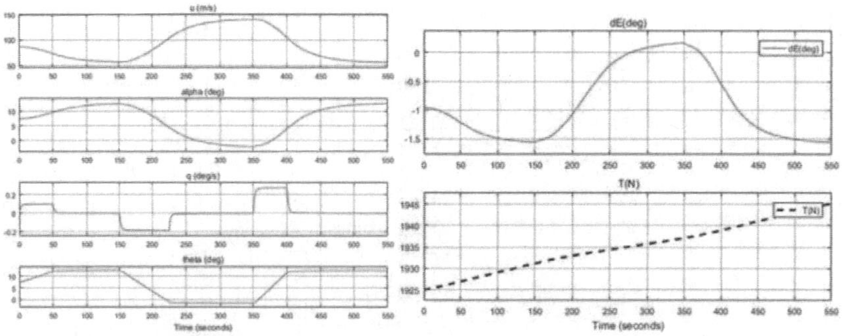


(a) (b)

**Figure 6.** Result of CAS state variable response (a) and control surface response (b) in longitudinal

It can be seen in Figure 5 even it doesn't succeed to track $\gamma$ control command during simulation, but $\theta$ control command can be tracked in 30 seconds after the change of angle. It is happened because the value of the state error for $\gamma$ variable in matrix Q is so small compared to the value of the state error for $\theta$ variable.

Then, the aircraft model's trajectory is shown in Figure 7.

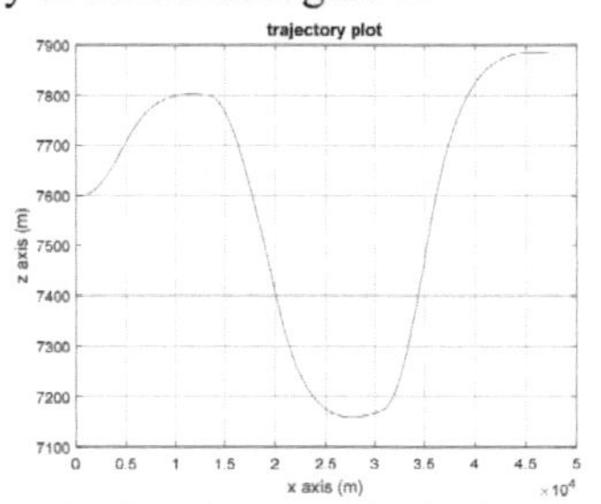


**Figure 7.** CAS trajectory plot in longitudinal.

Based on figure 7, aircraft's altitude changes during the changes of pitch angle. But, it can be assumed aircraft still in cruise flight because change of altitude isn't significant compared to change of distance.

## 7. Conclusion

CAS is succeed to track the control commands which are given both in longitudinal and lateral-directional even in longitudinal mode $\gamma$ variable is failed to track and in lateral-directional mode there is still oscillation in $\chi$ tracking result.

## References

[1] Kwon-Su Jeon, J.W. Lee, Y.H. Byun, Yung H. Yu 2007 Multidisciplinary UCAV system design and optimization using repetitive response surface enhancement technique, in: 3rd AIAA MDO Specialist Conference, Sheraton Waikiki, Honolulu,Hawaii

[2] Corless M 2006 *Course Notes*: *Linear Systems and Control* Purdue University Indiana p 364-6

[3] Nuriwati, Rika. 2018. *Sistem Kendali Pengatur Ketinggian Terbang Wahana UCAV dengan Total Energy Control System*. Bandung, Faculty of mechanical and aerospace engineering Institut Teknologi Bandung

[4] Ogata K 2010 *Modern Control Engineering fifth edition* New Jersey,Prentice Hall p 723-36, 793-95

IOP Conf. Series: Materials Science and Engineering **645** (2019) 012020 doi:10.1088/1757-899X/645/1/012020

# Development Flight Path Control for Unmanned Combat Aerial Vehicle (UCAV) Using Total Energy Control System (TECS)

**R Nuriwati and R A Sasongko**

Faculty of Mechanical and Aerospace Engineering, Institut Teknologi Bandung, Indonesia

Email: rikanuriwati@s.itb.ac.id

**Abstract**. UCAV (Unmanned Combat Aerial Vehicle) is a unmanned flying vehicle designed to carry out combat missions by carrying weapons. In the UCAV mission profile there is usually a cruise mission. In this cruise phase the vehicle must have a good control system and control altitude. In this paper, the UCAV altitude control system is designed to use the TECS (Total Energy Control System). TECS coordinates changes in speed and change in altitude by considering the total energy contained in the flying vehicle. The UCAV dynamic model that is used as the basis of the development is the fixed wing UCAV model which is equivalent to one of the UCAVs that are currently operating. As a comparison, a conventional control system (speed hold system and altitude hold system) was developed. The results showed that the altitude control system with TECS can adjust flight altitude and speed in a more coordinated manner, so that there is no change in the drastic and sudden variable values of flying when maneuvering. change in height begins.

## 1. Introduction

UCAV (Unmanned Combat Aerial Vehicle) is an unmanned flying vehicle designed to carry out combat missions or can be referred to as combat drones. The development of UCAV vehicles cannot be separated from the control design used to operate the vehicle in the air. One of the systems needed to operate the UCAV vehicle is a fly height control system used in the roaming flight phase. For the purpose of controlling this roaming phase, speed hold system can be used and the altitude hold system is operated together to maintain flight height and speed. This control design based on SISO (single input single output) still has some disadvantages in it, one of which is that there is no coordination between one system and another. Altitude hold system is used to adjust the height of the aircraft by giving the elevator deflection command, while the speed hold system is used to adjust the thrust needed by the aircraft to adjust the flight speed.

Both systems work without coordinating with each other, so that the response of the resulting plane to regulate elevators and thrust becomes out of sync. When one variable is controlled, the other variable will move away from the balance condition. For example, when a speed hold system is adjusting the speed with a fixed deflection of the elevator, crossing the vehicle will be affected. Likewise on the contrary, when the hold system's altitude is regulating the flight path with fixed thrust, there will be speed deviation.

One of the problem-solving strategies that has been used and tested is the Total Energy Control System (TECS). TECS is a control technique that uses the total mechanical energy of an aircraft to regulate flight altitude through the detection of total energy error and distribution energy errors. In the implementation of the regulation of flight height, changes in velocity are detected as changes in kinetic energy while changes in altitude (crossing flight) are detected as changes in the potential energy of the aircraft, as on Fig. 1. There is a coordination between changes in speed with changes in altitude that occur on the plane [1,2].

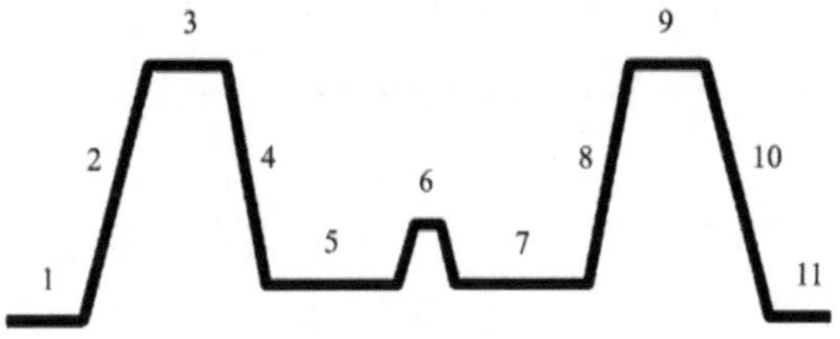


| No | Mission Segment | No | Mission Segment |
|---|---|---|---|
| 1. | Take-off | 7. | Dash out |
| 2. | Climb/Accelerate | 8. | Climb/decelerate |
| 3. | Cruise | 9. | Cruise |
| 4. | Descent/Accelerate | 10. | Descent/decelerate |
| 5. | Dash in | 11. | LAnding |
| 6. | Combat | | |

**Figure 1**. UCAV flight profile

Furthermore, a study and design of the Total Energy Control System will be conducted for the UCAV vehicle which has a mission profile as in Figure 1. The TECS model and analysis will be implemented to adjust the flight height on the cruise phase. The flight dynamics model as the basis of the control system design was built based on the motion equation of the vehicle in the roaming phase, where the aerodynamic parameters used were obtained using Digital DATCOM software [3]. Based on the flight dynamics model that has been built, further development of the fly height control system with TECS is developed. The resulting control system is then simulated and analyzed with MATLAB / Simulink software [4]. Analysis of the results obtained is complemented by a comparison between the TECS system and the fly height control system that uses a conventional approach, namely by using a combination of speed hold system and altitude hold system [5-7].

## 2. TECS

The concept of the Total Energy Control System (TECS) is built from the total formulation of the mechanical energy of an air vehicle by looking at the state of the vehicle represented in altitude, airspeed and flight path. If there is a change in the state of the vehicle, it will cause a total energy error and an energy distribution error. In this approach, to control the total energy error throttle is used while to control the energy distribution error between the flight path and airspeed an elevator is used.

The derivative of the TECS equation refers to reference [2,8], where the total energy of an air vehicle can be determined by adding up the potential energy and kinetic energy of the vehicle.

$$E = E_{KE} + E_{PE} = E = \frac{1}{2}mV^2 + mgh \tag{1}$$

With $E_{KE}$ is kinetic energy and $E_{PE}$ is potential energy. The equation has a constant variable in the form of mass, so that the equation can be simplified to:

$$E = mg\left(\frac{1}{2}\frac{V^2}{g} + h\right) \tag{2}$$

By using small-disturbance theory, relative to trim conditions, aircraft energy at steady flight conditions, can be expressed as:

$$E_0 + \Delta E = mg\left(\frac{1}{2}\frac{(V_0 + \Delta V)^2}{g} + (h_0 + \Delta h)\right) \tag{3}$$

Assuming ΔV is small $\Delta V^2 \approx 0$, so that:

$$\frac{d}{dt}(E_0 + \Delta E) = mg\frac{d}{dt}\left(\frac{1}{2}\left(\frac{V_0^2}{g} + \frac{2V_0\Delta V}{g}\right) + (h_0 + \Delta h)\right) \tag{4}$$

So that is obtained,

$$\Delta\dot{E} = mg\left(\frac{V_0\Delta\dot{V}}{g} + \Delta\dot{h}\right) \tag{5}$$

By dividing the above equation with the trim speed ($V_0$) obtained:

$$\frac{\Delta\dot{E}}{V_0} = mg\left(\frac{\Delta\dot{V}}{g} + \Delta\gamma\right) \tag{6}$$

With $\Delta\gamma \approx \frac{\Delta\dot{h}}{V_0}$, is the *flight path angle.*

Thus, it can be seen that for a certain speed the change in energy from an air vehicle depends on changes in speed and changes in the angle of flight.

If the change in the total energy of the air vehicle depends on the engine thrust given, then using the cruise flight equation is obtained:

$$m\Delta\dot{V} = \Delta T - \Delta D - mgsin\Delta\gamma \tag{7}$$

Assuming that Δγ is small so sinΔγ≈Δγ, modifying equation 7 will obtain the following relationship:

$$\Delta T = mg\left(\frac{\Delta\dot{V}}{g} + \Delta\gamma\right) + \Delta D = \frac{\Delta\dot{E}}{V_0} + \Delta D = \frac{\Delta\dot{E}_s}{V_0}mg + \Delta D \tag{8}$$

Assuming that in the steady flight conditions change the drag is small, the relationship is obtained:

$$\Delta T \propto \left(\Delta\gamma + \frac{\Delta\dot{V}}{g}\right) \tag{9}$$

Then the relationship between the total specific energy rate ($\Delta\dot{E}_s$) is controlled by the throttle and the conditions of the flight vehicle are as follows:

(11)

$$\Delta T \propto \Delta\gamma + \frac{\Delta a_{long}}{g} = \Delta\dot{E}_s \tag{12}$$

$$\Delta\dot{E}_s = \dot{E}_s - \dot{E}_{s0}$$

And for the specific energy rate distribution error ($\Delta\dot{L}_s$) referring to reference [8,9], it is a potential energy minus kinetic energy. Where the elevator can be used as a transformation of potential energy into kinetic energy and vice versa. So that $\Delta\dot{L}_s$ is stated as:

$$\Delta\dot{L}_s = -\Delta\gamma + \frac{\Delta a_{long}}{g} \tag{13}$$

Where $\Delta a_{long}$ is acceleration in the plane of symmetry of the vehicle or in the longitudinal dimension.

## 3. Aircraft Model

UCAV used has a fixed-wing configuration with Y-tail. And it is equipped with great power compared to similar aircraft, making it possible to carry more payloads and be able to cruise at higher speeds. The following UCAV specifications are used, based on references [10,11], refer to Table 1.

**Table 1.** Aircraft specification

| Specification | value | unit |
|---|---|---|
| **Mass** | 4760 | kg |
| **Wing span** | 20 | m |
| **Wing area** | 17.7 | $m^2$ |
| **HTP area** | 6.05 | $m^2$ |
| **VTP area** | 1.45 | $m^2$ |
| **Engine max power** | 115 | hp |
| **Propeller diameter** | 2.8 | m |

## 4. TECS Model

TECS is used to control the vehicle in the longitudinal dimension, more precisely to control the altitude of the vehicle. In accordance with the equation stated in section II, the parameters used are the specific energy distribution rate ($\dot{L}_S$) and the total energy specific rate ($\dot{E}_S$). The following is the TECS block diagram used, as shown in Fig. 2.

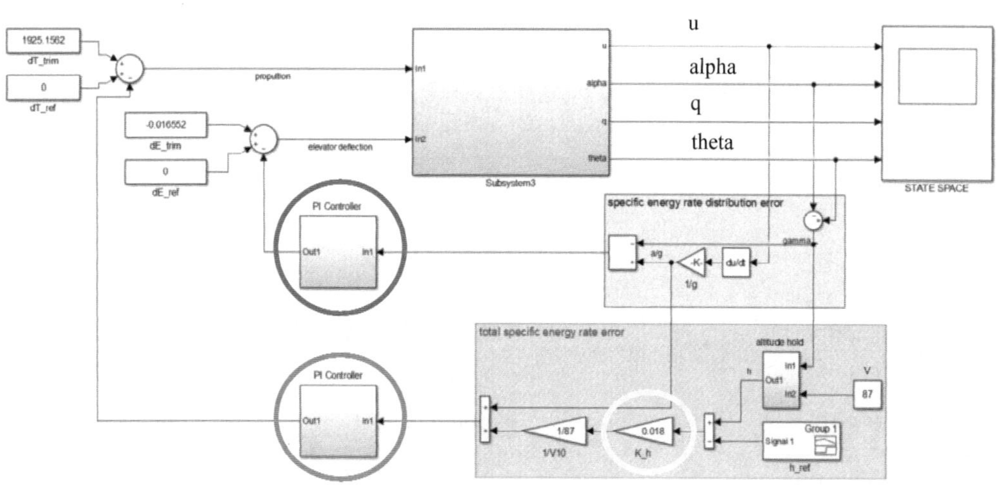


**Figure 2**. TECS linear model

The inner loop is an energy distribution error ($\dot{L}_S$) which is controlled using an elevator. Whereas the outer loop is the total energy error ($\dot{E}_S$) which is controlled by thrust, $\dot{L}_S$ and $\dot{E}_S$ are directed to zero. Seen in Figure 2, the TECS block design uses PI gain (manual tuning) for the energy distribution error circled in red. For total energy error, the Kh gain (yellow circle) is used to get the corresponding h˙and PI gain (blue circle).

## 5. Results and Discussions

### *5.1. Effect of H gain*

In designing TECS it takes several gains to help the system achieve the expected conditions. One of these gains is the gain of Kh, following the effect of the Kh gain on the vehicle response, as shown in Table 2 and Fig. 3.

IOP Conf. Series: Materials Science and Engineering **645** (2019) 012020 doi:10.1088/1757-899X/645/1/012020

**Table 2.** Kh gain variation

| Gain | Gain Variation |
|---|---|
| Kh1 | 0.015 |
| Kh2 | 0.018 |
| Kh3 | 0.021 |
| Kh4 | 0.024 |

(a) (b) (c)

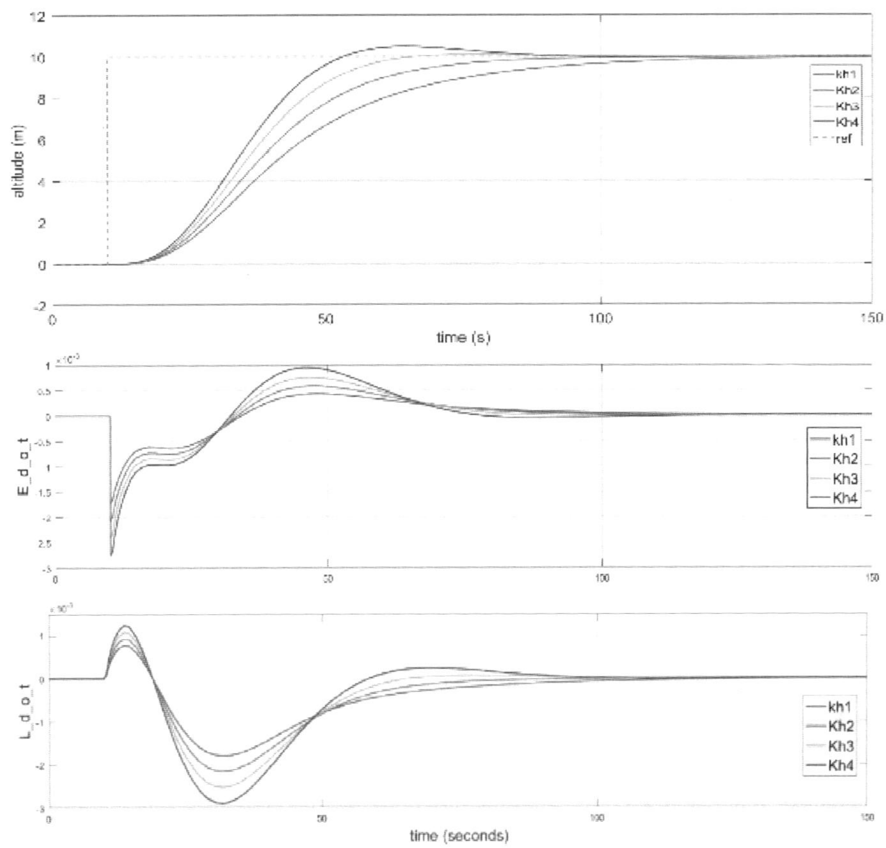


**Figure 3.** Response of (a) altitude, (b) total energy error, and (c) energy distribution error on Kh gain variation.

This gain greatly affects the rate of climb of the vehicle, it can be seen in Figure 3(a) that the greater the gain value of Kh, the speed of the vehicle to reach the intended height is faster. Kh1 has the smallest rate of climb while Kh4 has the largest rate of climb. But if the gain used is too large, overshoot will occur. The fast and slow vehicle response can be verified by the total energy error ($\dot{E}_s$) and energy distribution error ($\dot{L}_s$) responses that occur in each gain, see Figures 3(b) and 3(c).

The faster the vehicle goes to the height that is ordered, ($\dot{L}_s$) and ($\dot{E}_s$) are also getting bigger. Because rides require greater energy and also greater energy distribution.

### 5.2. *Effect of P Gain*

The following will discuss the effect of changes in P (proportional) gain on the energy distribution error with P gain variation used in Table 3 and Fig. 4.

**Table 3.** P (Proportional) gain variation

| Gain | Proportional | Integral |
|---|---|---|
| kL1 | -0.005 | 0.0036 |
| kL2 | 0 | 0.0036 |
| kL3 | 0.01 | 0.0036 |
| kL4 | 0.02 | 0.0036 |
| kL5 | 0.03 | 0.0036 |

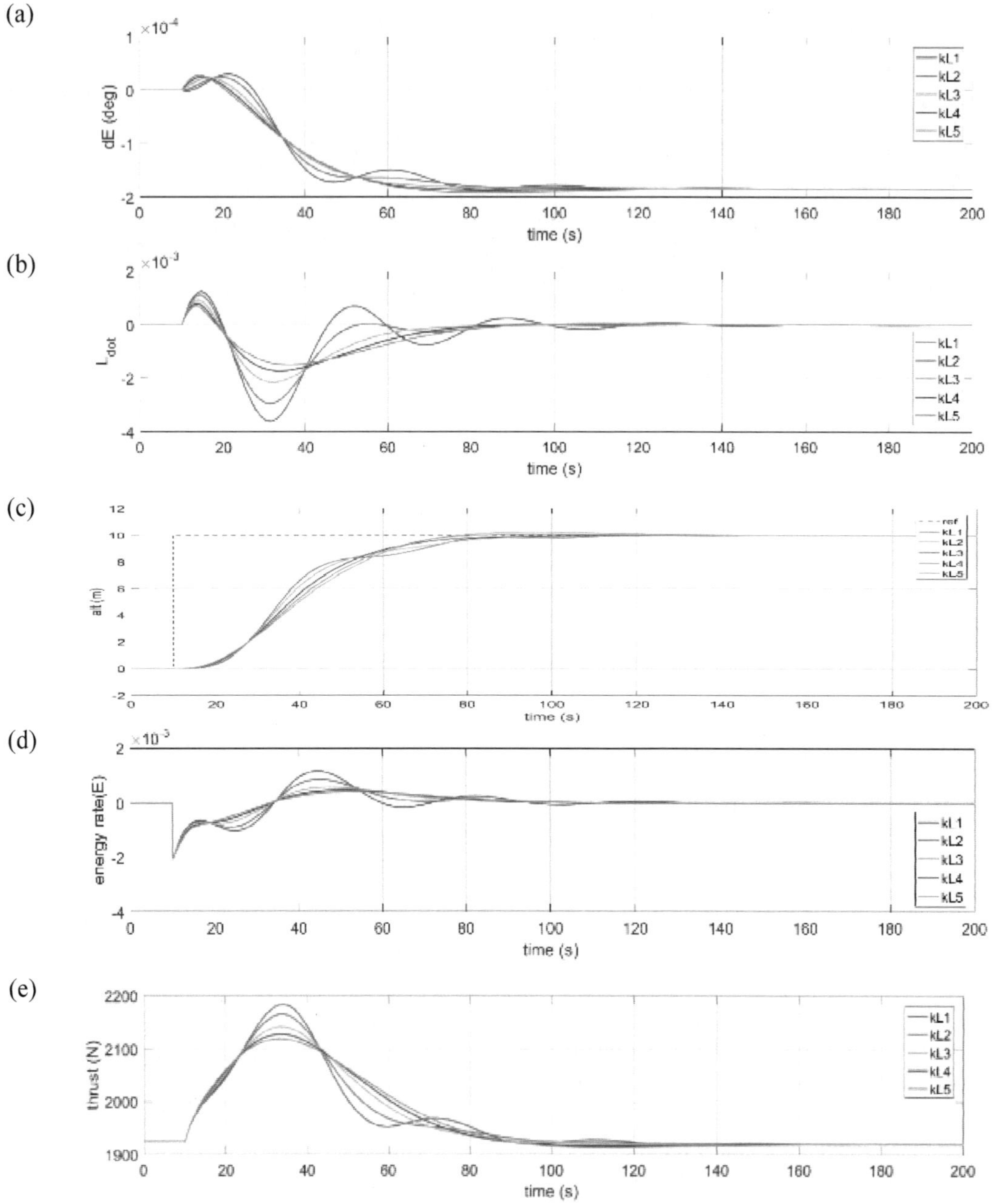


**Figure 4.** The effect of P gain variation: (a) elevator input, (b) energy distribution error, (c) altitude response, (d) energy rate response, and (e) thrust response.

Figure 4(a), the P (proportional) gain directly affects the elevator input. At 10-30 seconds it appears that kL1 has a smaller elevator deflection command compared to kL5. So that kL1 has a lower rate of climb compared to kL5 as seen in Figure 4(c). Changes in the negative elevator deflection indicate the plane is moving pitch up, the negative the elevator is the faster the altitude changes. In Figure 4(b) it can be seen that the greater the gain P, the smoother the energy distribution of the system with the change in the deflection of the elevator which tends to be sloping and constant. This results in a smaller change in the energy error of the system so that smaller delta thrust is also needed, see in Figures 4(d) and 4(e).

### 5.3. *Effect of I Gain*

The following will discuss the effect of changes in gain I (integral) in the energy distribution error, as presented in Table 4 and Fig. 5.

**Table 4.** Integral (I) gain variation

| Gain | Proportional | Integral |
|---|---|---|
| kL1 | 0.01 | 0.002 |
| kL2 | 0.01 | 0.003 |
| kL3 | 0.01 | 0.0036 |
| kL4 | 0.01 | 0.004 |
| kL5 | 0.01 | 0.005 |

In Figure 5(c) it can be seen that the greater the gain I, the system response will be smoother and the system will be faster to the equilibrium point. Changes in gain I almost have the same effect as the gain P. However, on gain I the changes that occur will affect vehicle trim conditions. Seen in Figure 5(c), the smaller the gain value I (kL1) in this state there is an increase in altitude which is quite dipping and there is a small increase in elevator deflection from the trim condition, see Figure 5(d). So, the response gain kL1 experiences a high climb rate and experiencing an overshoot which increases the total energy rate so that the thrust needed is also greater. Broadly speaking, the gain used to control the elevator through the energy distribution error ($\dot{L}_s$) affects the total energy conversion given by thrust. The more negative the deflection of the elevator will increase the height and the speed will decrease, and vice versa.

### 5.4. *TECS vs Conventional*

To compare the TECS system with conventional systems, several conventional combinations of gain are performed to get a response similar to the TECS response [8]. A combination of gain k1, k2, k3, k4 and k5 is used. Then, a combination of k1 gain is chosen because when viewed from the response, the k1 gain has a rate of climb similar to TECS as seen in Figure 6. However, when the system is given an altitude command, there is a delay in the TECS while in the conventional system the command is immediately executed, Fig. 6(a). Based on the input thrust required in Figure 6(b) it is seen that conventional systems require additional thrust to be slightly lower when compared to TECS. However, the conventional system of thrust changes that occur immediately increases shortly after being given an order to change altitude, while in TECS this does not occur. In Figure 6(c), the change in elevator deflection in a conventional system is larger and immediately returns to the original trim condition, while in TECS the deflection of the elevator is smaller and will be constant in the new trim condition.

Figure 6(d), at the speed change, the TECS will increase and arrive at the new trim condition whereas in the conventional system there is a decrease in speed, but after the height is reached the speed hold system returns the speed to the trim condition.

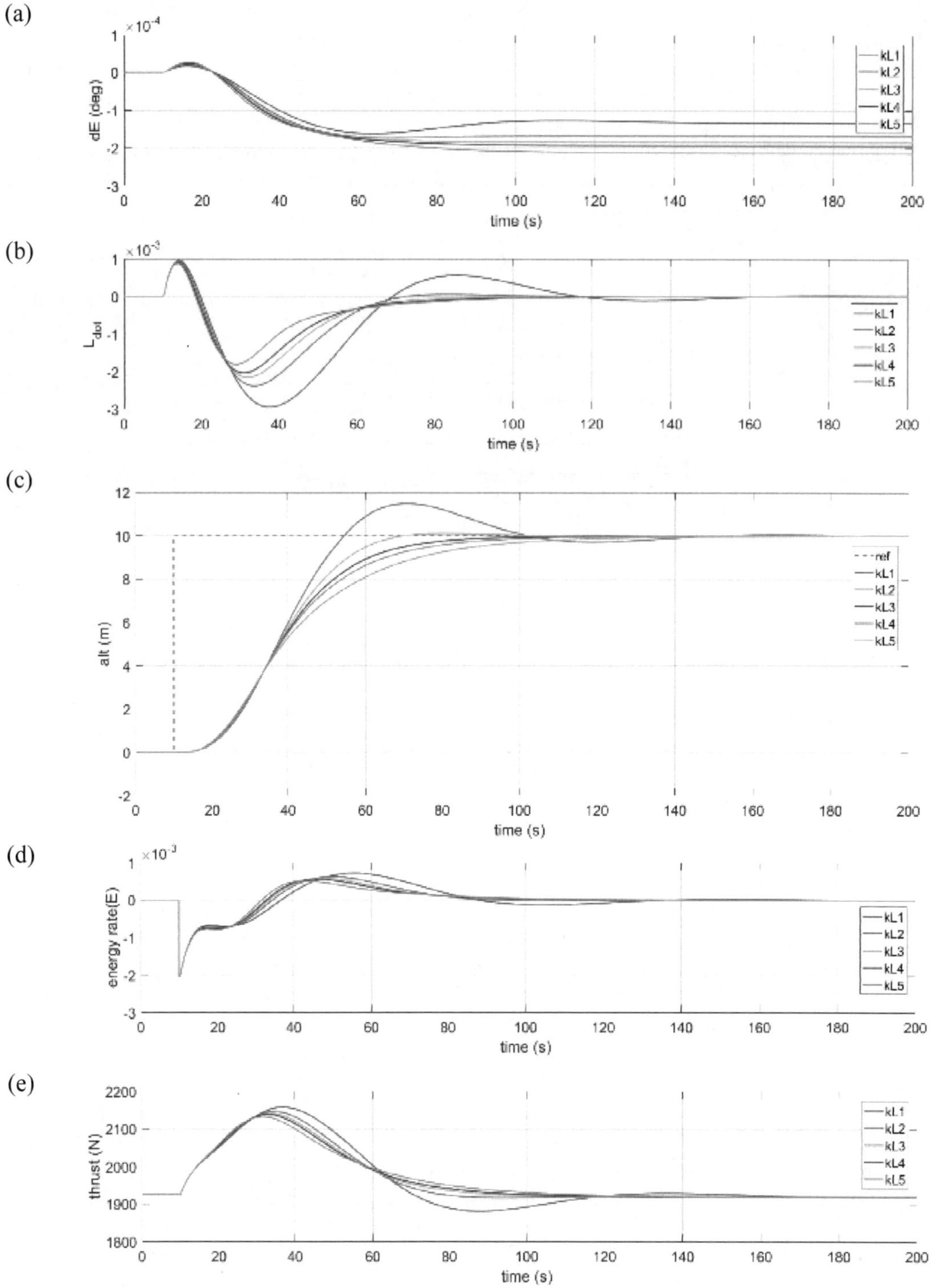


**Figure 5.** The effect of I gain variation: (a) elevator input, (b) energy distribution error, (c) altitude response, (d) energy rate response, and (e) thrust response.

IOP Conf. Series: Materials Science and Engineering **645** (2019) 012020 doi:10.1088/1757-899X/645/1/012020

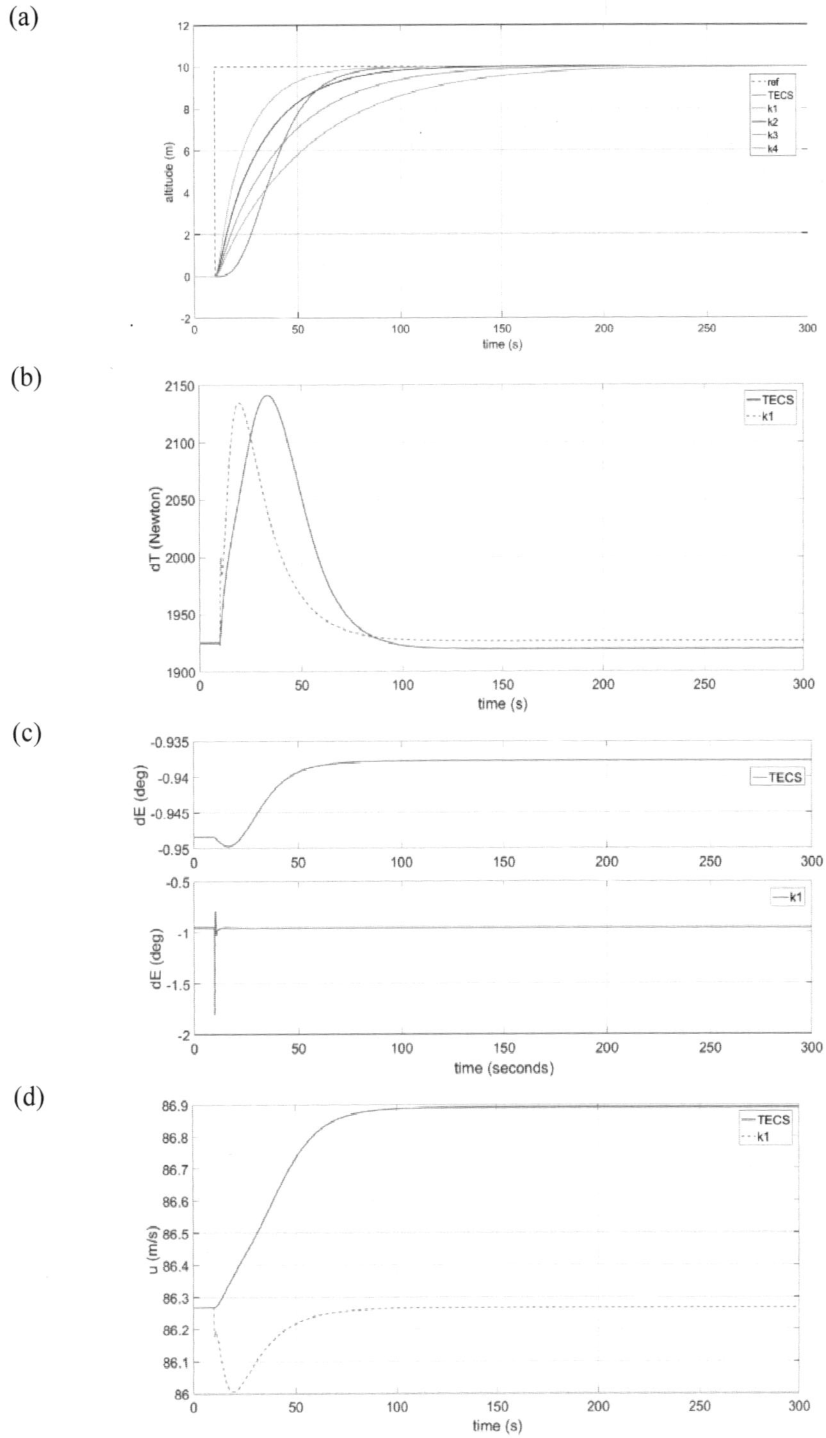


**Figure 6.** Comparison of conventional system vs TECS: (a) altitude hold, (b) delta thrust response, (c) elevator deflection, and (d) velocity response.

## 6. Conclusion

The development and analysis of height control with TECS can be obtained the TECS simulations on non-linear systems and linear systems have values that are not much different from the same gain range. TECS and conventional systems have advantages and disadvantages, in general the TECS system will have a new system trim condition. While the conventional system will go to a trim value. On the TECS form the response of the system to the height that is ordered depends on the gain of Kh. In the TECS the system energy distribution can be set using PI gain on the ($\dot{L}_s$) control. TECS provides a more gradual / smooth control variable response and flight variables, there is no large variable value spike at the beginning of the maneuver.

## References

[1] H. Muhammad dan Y. I. Jenie, Diktat Kuliah AE3231 Dinamika Terbang, Bandung: Faculty of Mechanical and Aerospace Engineering, Bandung Institut of Technology, 2011

[2] Lai, Ying-Chic; Ting, Wen Ong. 2016. Design and Implementation of an Optimal Energy Control System for Fixed-Wing Unmanned Aerial Vehicles. Taiwan, 2016

[3] McDonnell Douglas Astronautics Company St. Louis Division. 1979. The USAF Stability and Control DATCOM Volume I, User Manual. Missouri, USA: McDonnell Douglas Astronautics Company.

[4] MathWorks. 2015. Simulink User's Guide. USA : TheMathWorks, Inc.

[5] Ogata, Katsuhiko. 2010. Modern Control Engineering. New Jersey : Pearson Education, Inc.

[6] Sasongko, Rianto Adhy.2015. Modul Kuliah Kendali Terbang Otomatik. Bandung: Faculty of Mechanical and Aerospace Engineering, Bandung Institut of Technology.

[7] https://www.aopa.org/news-and-media/all-news/2015/august/flight-training-magazine/energy-management diakses 16 September 2018

[8] Arygle, Matthew E.; Beard, Randal W. Nonlinear Total Energy Control for the Longitudinal Dynamics of an Aircraf

[9] Brigido-González, J.; Rodríguez-Cortés, H. Experimental validation of an adaptive total energy control system strategy for the longitudinal dynamics of a fixed-wing aircraft. J. Aerosp. Eng. 2015, 29, 04015024

[10] https://www.washingtonpost.com/sf/investigative/2014/06/20/when-drones-fall-from-the-sky/?utm_term=.932c78e827bc diakses pada 2 September 2018

[11] Headquarters Allied Air Command Public Affairs Office. 2015. General Atomic MQ-9 Reaper. Nato Media Information centre Trident Juncture 2015.

# Design and Performance Analysis of Low Reynolds Number Propeller using Analytical Methods by Varying Blades Alpha Design

**M D F Haidar[1], M A Moelyadi[1], and F Hartono[1,*]**

Department of Aerospace Engineering, Institute of Technolgy Bandung, Indonesia

[*] Email: firman_hart@ae.itb.ac.id

**Abstract.** In the past few years, the usage of Unmanned Aerial Vehicles (UAVs) aircraft has increased significantly, but the need for UAVs with higher performance is increasing also. Therefore, all systems of the UAVs must be developed to have high efficiency, including the propulsion system. It is necessary to perform some design and analysis tasks on low Reynolds number propeller, which is a commonly used propulsion system on UAVs, to have higher efficiency and improve the overall UAVs performance. In this paper, low Reynolds number propellers are designed using the Larrabee Design analytical method by referring to the "Master AirScrew 10x7E" propeller. Meanwhile, performance of the designed propellers is analysed by using analytical method of Inversing Larrabee Design method. At the design stage of low Reynolds number propeller, 42 variations of alpha design ($\alpha_{des}$) was inputted along the blades with uniform, linear, and quadratic distributions. Furthermore, the best design from these 42 variations was selected by performing two-stage scoring. At first stage scoring, 42 designs are selected to obtain 5 designs with the highest on-design efficiency. Then, performance of these 5 selected designs are analysed analytically along a certain range of advance ratio. Finally, second stage scoring is performed to obtain 1 design with the best overall performance.

**Keywords**: low Reynolds number propeller, efficiency, propeller design, alpha design, Larrabee method.

## 1. Introduction

The usage of Unmanned Aerial Vehicles (UAVs) aircraft have boomed in the past few years, both for military or commercial application. The need for UAVs with higher performance is also increasing. Therefore, performance of all systems of the UAVs must be increased to achieve UAV with higher efficiency, including the propulsion system. Propulsion system that is commonly used on UAVs aircraft is non-conventional propeller which has small diameter, operating at lower Reynolds number (based on chord at 0.75 radius location) , low axial speed, and high revolution per minute (RPM). The researches on this type of propeller are relatively rare. Therefore, it is very important to design and analyse the performance of the low Reynolds number propeller to have higher efficiency, so that the UAV aircraft has reliable performance.

There are several common methods that can be used to perform design and analysis tasks on propeller performance. These methods are analytical methods, numerical simulations using computational tools, and experimental methods. In analytical methods, design and performance analysis of propeller is performed based on the development of various propeller theories, such as vortex theory, blade element theory, momentum theory, and a combination thereof. In numerical simulation methods, propeller performance analysis is carried out by utilizing the Reynold Average Navier Stoke (RANS) equation to solve physical problem of propeller phenomena using Computational Fluid Dynamics (CFD). Whereas in the experimental methods, the propeller performance analysis is carried out by testing the propeller using wind tunnel, both in static conditions and along a certain range of advance ratio.

This research focuses on design and performance analysis of low Reynolds number propeller using analytical method by varying alpha design ($\alpha_{des}$). Several studies have been carried out by other researchers previously related to this research. Brand et al. [1], have carried out experimental studies using wind tunnel to determine the performance of 79 low Reynolds number propellers ranging in diameter 9 to 11 inch. The experiment was conducted to see the impact of Reynolds numbers and advance ratio on propeller performance. The study shows that propeller performance is increased with increasing of Reynolds number. Similar research was conducted by Merchant et al. [2], who performed experimental studies using wind tunnel to determine performance of low Reynolds number propellers at ranges of Reynolds number from 30,000 to 300,000. Kutty, et al. [3] performed a numerical CFD simulation to predict the performance of small-sized propeller 'APC Slow Flyer' using ANSYS Fluent solver. The results of numerical simulations were then compared with the 'APC Slow Flyer' propeller experimental data. The results show that the propeller performance data that were predicted using numerical simulations are close enough to the propeller performance data that were obtained experimentally. Kumar et al. [4] also conducted an analysis of 9 inches diameter propeller using the numerical CFD simulation method. Before analyzing with CFD, CAD data of propeller was obtained using the reverse engineering method. The thrust value from numerical simulation method was then compared to the thrust value from the analytical calculation. The difference in the value of thrust resulting from these two methods is 27.12%. Research that is related to the design and analysis of the performance of low Reynolds number propellers still requires further development.

## 2. Theoretical Formulation

### *2.1. Larrabee Design Method*

Larrabee design is an analytical method that is used to design a propeller. This method is based on momentum theory, blade element theory, and vortex theory. Larrabee design method calculates the behavior of a propeller by dividing the propeller blades into small parts. Each part of this blade acts as an airfoil that produces two-dimensional aerodynamic forces. This method requires power or thrust coefficient desired ($C_{P_{des}}$ or $C_{T_{des}}$), propeller diameter ($D$), hub to tip ratio ($HTR$), number of propeller blades ($B$), operating condition of propeller ($V_0$, $RPM, h, \rho, \mu$), alpha design ($\alpha_{des}$) and airfoil data ($c_l$, $c_d$) as inputs. Equation of thrust and power on each blade element in this method are presented in Eq. (1) and (2),

$$\frac{dT_c}{d\xi} = 4\xi\zeta F \frac{x^2(1+\zeta)}{[(1+\zeta)^2+x^2]}\left\{1-\frac{1}{2}\zeta\frac{1+\zeta}{[(1+\zeta)^2+x^2]}\right\} \left\{1-\frac{D}{L}\frac{1}{x}\frac{2[(1+\zeta)^2+x^2]+\zeta x^2}{2[(1+\zeta)^2+x^2]-(1+\zeta)\zeta}\right\} \tag{1}$$

$$\frac{dP_c}{d\xi} = 4\xi\zeta F \frac{x^2(1+\zeta)}{[(1+\zeta)^2+x^2]}\left\{1+\frac{1}{2}\zeta\frac{x^2}{[(1+\zeta)^2+x^2]}\right\} \left\{1+\frac{D}{L}x\frac{2[(1+\zeta)^2+x^2]-(1+\zeta)\zeta}{2[(1+\zeta)^2+x^2]+\zeta x^2}\right\} \tag{2}$$

Based on this method, propeller performance at on-design condition can be obtained.

### *2.2. Inversing Larrabee Design Method*

Inversing Larrabee design method is an analytical method to analyse performance of propeller. This method requires propeller diameter ($D$), hub to tip ratio ($HTR$), number of propeller blades ($B$), chord distribution ($c/R$), twist distribution ($\theta$), operating condition of propeller ($V_0,, h, \rho, \mu$), and airfoil data ($c_l, c_d$) as inputs. Propeller performance analysis is performed by comparing the thrust value from general momentum theory and blade element theory iteratively. Iteration is executed continuously until thrust that is calculated from these 2 theories generates the same value. Thrust on each propeller blade element based on the general momentum theory is obtained by using Eq. (1). Whereas based on blade element theory, thrust produced by each blade element is obtained by using the following equation,

$$\frac{dT_c}{d\xi} = \frac{B}{\pi}\left(\frac{W}{V}\right)^2 \frac{c}{R}[C_l \cos \emptyset_a - C_d \sin \emptyset_a] \tag{3}$$

Propeller performance is commonly expressed as non-dimensional quantities. The general parameters used to analyse propeller performance are thrust coefficient, power coefficient, and propeller efficiency, which are obtained using the following equation,

$$C_T = \frac{T}{\rho n^2 D^4} \tag{4}$$

$$C_P = \frac{P}{\rho n^3 D^5} \tag{5}$$

$$\eta = J\frac{C_T}{C_P} \tag{6}$$

while $n$ states number of propeller revolutions per second and $J$ is an advance ratio which states the relationship between axial speed, propeller rotational speed, and propeller diameter with the following equation,

$$n = \frac{RPM}{60} \tag{7}$$

$$J = \frac{V_0}{nD} \tag{8}$$

### *2.3. Scoring*

Scoring is a method used to select the best design from the overall design by giving a score to each design. The design with the highest score is considered as the best design. In this study, the best design selection was obtained by conducting two scoring stages which will be explained as follows,

*First stage scoring.* On this stage, scoring is carried out on the design advance ratio ($J$=0.643) by comparing the performance of the designed propeller with the performance of the 'Master AirScrew 10x7E' reference propeller. The formula used on first stage scoring is as follows,

$$S_1 = \frac{\eta_{design}\big|_{J_{design}}}{10}\left(f_P\big|_{J_{design}} + 4f_T\big|_{J_{design}} + 5f_\eta\big|_{J_{design}}\right) \tag{9}$$

with,

- $$f_P\big|_{J_{design}} = \begin{cases} 0, C_{P_{design}}\big|_{J_{design}} - C_{P_{Master\ Airscrew}}\big|_{J_{design}} \geq 0 \\ 1, C_{P_{design}}\big|_{J_{design}} - C_{P_{Master\ Airscrew}}\big|_{J_{design}} < 0 \end{cases}$$

- $f_T|_{J_{design}} = \begin{cases} 0, C_{T_{design}}\Big|_{J_{design}} - C_{T_{Master\ Airscrew}}\Big|_{J_{design}} \le 0 \\ 1, C_{T_{design}}\Big|_{J_{design}} - C_{T_{Master\ Airscrew}}\Big|_{J_{design}} > 0 \end{cases}$

- $f_\eta\Big|_{J_{design}} = \begin{cases} 0, \eta_{design}\Big|_{J_{design}} - \eta_{Master\ Airscrew}|_{J_{design}} \le 0 \\ 1, \eta_{design}\Big|_{J_{design}} - \eta_{Master\ Airscrew}|_{J_{design}} > 0 \end{cases}$

From this stage, 5 designs with the highest score is chosen to be selected in scoring stage 2.

*Second stage scoring.* Second stage scoring is carried out by comparing the performance of designed propeller and performance of the 'Master AirScrew 10x7E' reference propeller at each values of advance ratio ranging from 0.147 to 0.712. That advance ratio range is chosen based on the availability of experimental performance data of the 'Master AirScrew 10x7E' propeller. Before second stage scoring is executed, propeller performance analysis was carried out on 5 selected designs at ranges of advance ratio from 0.147 to 0.712 using inversing Larrabee method. The formula used for Second stage scoring is as follows,

$$S_2 = \frac{1}{\sum_{i=1}^{9} k_i} \sum_{i=1}^{9} \left[ k_i \frac{\eta_{design}\big|_{J_i}}{10} \left( f_P\big|_{J_i} + 4 f_T|_{J_i} + 5 f_\eta\big|_{J_i} \right) \right] \quad (10)$$

with,

- $f_P\Big|_{J_i} = \begin{cases} 0, C_{P_{design}}\Big|_{J_i} - C_{P_{Master\ Airscrew}}\Big|_{J_i} \ge 0 \\ 1, C_{P_{design}}\Big|_{J_i} - C_{P_{Master\ Airscrew}}\Big|_{J_i} < 0 \end{cases}$

- $f_T|_{J_i} = \begin{cases} 0, C_{T_{design}}\Big|_{J_i} - C_{T_{Master\ Airscrew}}\Big|_{J_i} \le 0 \\ 1, C_{T_{design}}\Big|_{J_i} - C_{T_{Master\ Airscrew}}\Big|_{J_i} > 0 \end{cases}$

- $f_\eta\Big|_{J_i} = \begin{cases} 0, \eta_{design}\Big|_{J_i} - \eta_{Master\ Airscrew}|_{J_i} \le 0 \\ 1, \eta_{design}\Big|_{J_i} - \eta_{Master\ Airscrew}|_{J_i} > 0 \end{cases}$

and,

**Table 1.** The Value of Multiplier Factor at Each Advance Ratio

| *i* | *J* | *k* |
|---|---|---|
| **1** | 0.147 | 12.5 |
| **2** | 0.218 | 25 |
| **3** | 0.286 | 37.5 |
| **4** | 0.360 | 50 |
| **5** | 0.423 | 62.5 |
| **6** | 0.504 | 75 |
| **7** | 0.575 | 87.5 |
| **8** | 0.643 | 100 |
| **9** | 0.712 | 87.5 |

From Table 1, it can be seen that multiplier factor ($k$) is set so that it has gradual values along the ranges of advance ratio which are analyzed in this scoring stage. So, propeller performance on the advance ratio value which is closer to on-design operating conditions ($J$ = 0.643) has a greater effect on scoring compared to the advance ratio value which is further from on-design operating conditions. For example, it can be seen that multiplier factor at $J$ = 0.147 is smaller than multiplier factor at $J$ = 0.575. Meanwhile, the multiplier factor on the on-design condition ($J$ = 0.643) has the biggest value.

## 3. Low Reynolds Number Propeller Design

In this research, Design Requirement and Objectives (DRO) of propeller refers to the geometrical characteristics of low Reynolds number propeller 'Master Air Screw 10x7 E' and its sea-level performance at 4008 RPM that was obtained through experimental results conducted by Brandt [5]. Fig. 1 show experimental results of sea level performance of 'Master Air Screw 10x7 E' propeller at 4008 RPM,

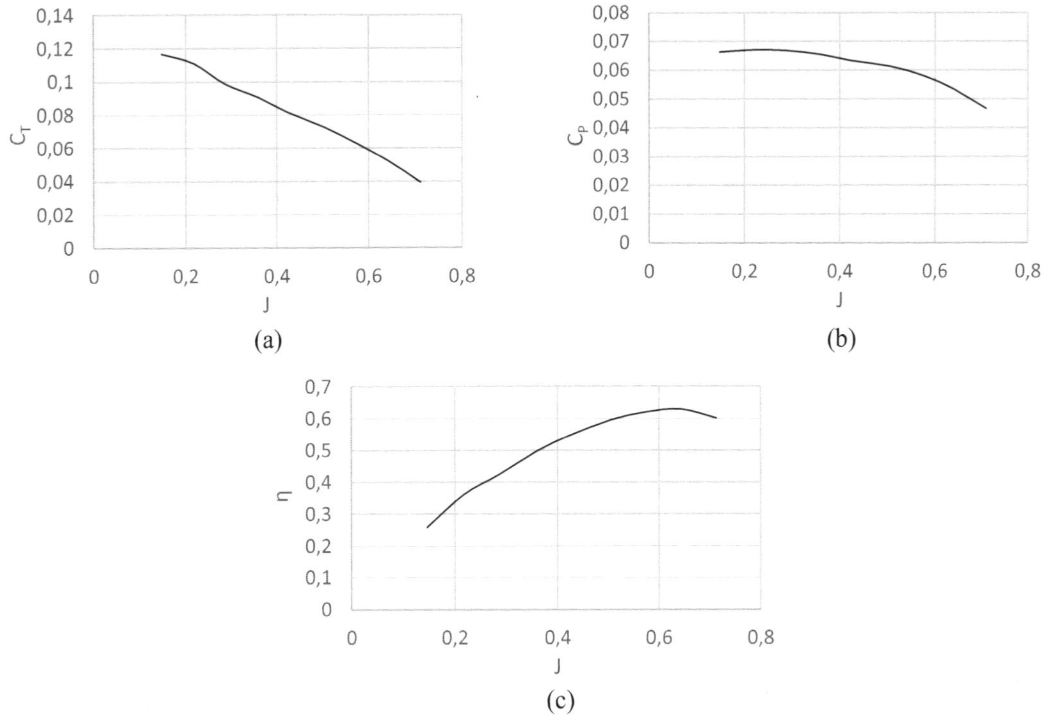


**Figure 1**. Sea-Level Performance of 'Master Air Screw 10x7E' Propeller at 4008 RPM, (a) $C_T$ vs $J$, (b) $C_P$ vs $J$, and (c) $\eta$ vs $J$

From Fig. 1(c), it can be seen that the maximum efficiency of the "Master AirScrew 10x7E" propeller based on experimental results at 4008 RPM is obtained at the advance ratio of 0.643 with an efficiency value of 0.629. On that advance ratio, it can be obtained propeller thrust and power coefficient value based on Fig. 1(a) and (b) respectively is 0.052 and 0.053. Then using Eq. (8), the axial velocity of the propeller on that advance ratio is 10.91 m/s. Then, it is obtained DRO specifications that are used in this study are as follows,

**Table 2**. Design Requirement and Objectives (DRO) Specification of Low Reynolds Number Propeller

| **Design Requirement** | |
|---|---|
| **Axial velocity ($V_0$)** | 10.91 m/s |
| **Rotational speed ($RPM$)** | 4008 |
| **Diameter ($D$)** | 10 in |
| **Number of blades ($B$)** | 2 |

IOP Conf. Series: Materials Science and Engineering **645** (2019) 012021 doi:10.1088/1757-899X/645/1/012021

| | |
|---|---|
| ***Hub to tip ratio (HTR)*** | 0.15 |
| **Altitude ($h$)** | 0 m (*sea-level*) |
| **Advance Ratio ($J$)** | 0.643 |
| **Design Objectives** | |
| **Power coefficient desired ($C_{P_{des}}$)** | 0.053 |

The type airfoil that is used by the propeller design in this study is NACA 4412.

## 4. Study of Alpha Design ($\alpha_{des}$) Variations

In this research, propeller design are varied by giving 42 different $\alpha_{des}$ inputs. From these 42 $\alpha_{des}$ inputs, there are 6 inputs with $\alpha_{des}$ characteristics that are uniform along the blade span, 8 inputs with the characteristics of $\alpha_{des}$ are linear distribution, and 28 inputs with $\alpha_{des}$ characteristics are quadratic distribution. All variations of $\alpha_{des}$ that were inputted in this research are shown in Fig. 2 below.

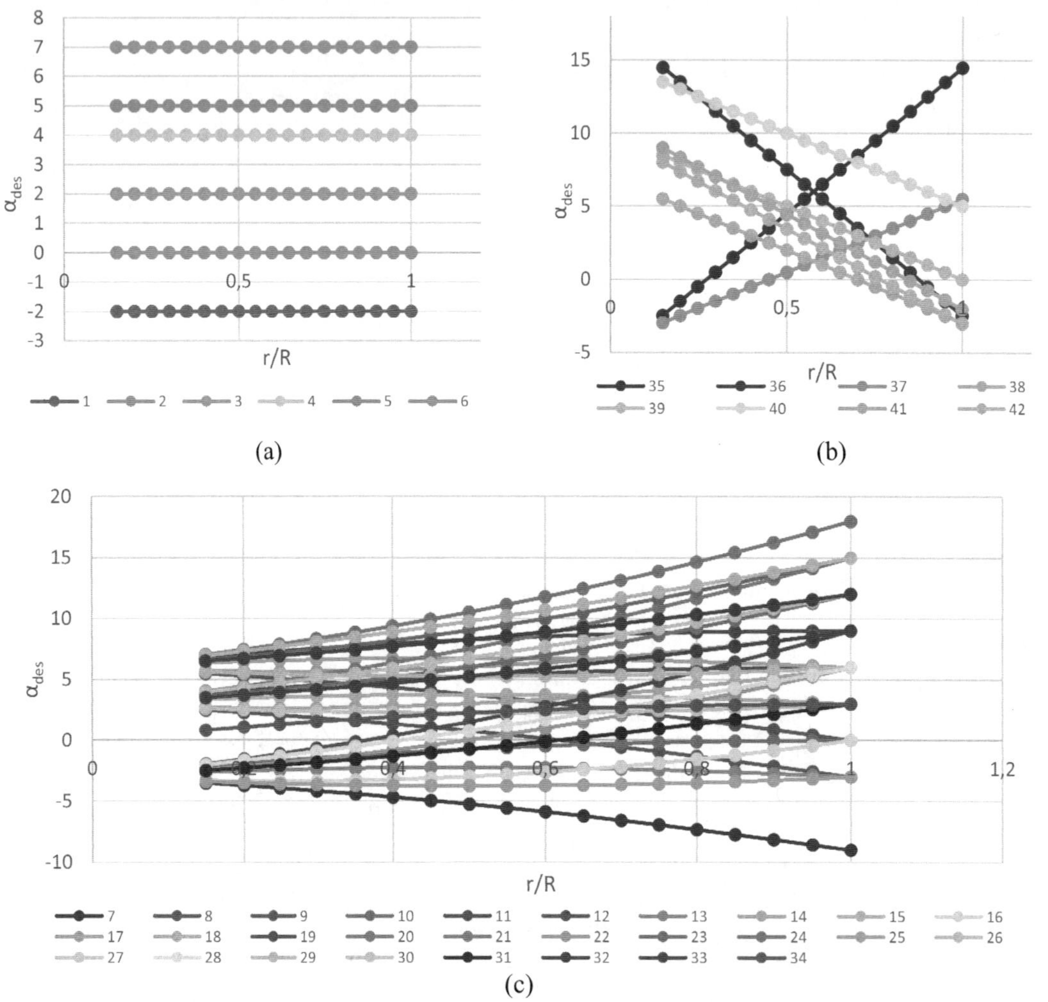


**Figure 2**. Alpha Design Along the Blade Span, (a) Uniform Distribution, (b) Linear Distribution, (c) Quadratic Distribution

## 5. Results

The on-design performances of propellers are obtaining which include thrust coefficient, power coefficient, and efficiency of each design as follows.

**Table 3**. On-Design Performance of All Designed Propellers

| Design | $C_T$ | $C_P$ | $\eta$ | Design | $C_T$ | $C_P$ | $\eta$ | Design | $C_T$ | $C_P$ | $\eta$ |
|---|---|---|---|---|---|---|---|---|---|---|---|
| 1 | 0.060 | 0.053 | 0.721 | 15 | 0.065 | 0.053 | 0.783 | 29 | 0.054 | 0.053 | 0.651 |
| 2 | 0.063 | 0.053 | 0.766 | 16 | - | - | - | 30 | 0.061 | 0.053 | 0.733 |
| 3 | 0.065 | 0.053 | 0.782 | 17 | 0.064 | 0.053 | 0.768 | 31 | 0.063 | 0.053 | 0.763 |
| 4 | 0.065 | 0.053 | 0.783 | 18 | 0.065 | 0.053 | 0.782 | 32 | 0.059 | 0.053 | 0.708 |
| 5 | 0.065 | 0.053 | 0.780 | 19 | 0.064 | 0.053 | 0.769 | 33 | 0.064 | 0.053 | 0.769 |
| 6 | 0.064 | 0.053 | 0.769 | 20 | 0.049 | 0.053 | 0.592 | 34 | 0.065 | 0.053 | 0.782 |
| 7 | - | - | - | 21 | 0.057 | 0.053 | 0.684 | 35 | 0.058 | 0.053 | 0.707 |
| 8 | 0.065 | 0.053 | 0.782 | 22 | 0.064 | 0.053 | 0.768 | 36 | 0.063 | 0.053 | 0.760 |
| 9 | 0.061 | 0.053 | 0.740 | 23 | 0.055 | 0.053 | 0.665 | 37 | 0.064 | 0.053 | 0.769 |
| 10 | 0.062 | 0.053 | 0.755 | 24 | 0.061 | 0.053 | 0.733 | 38 | 0.062 | 0.053 | 0.748 |
| 11 | 0.061 | 0.053 | 0.734 | 25 | - | - | - | 39 | 0.065 | 0.053 | 0.782 |
| 12 | 0.064 | 0.053 | 0.776 | 26 | 0.064 | 0.053 | 0.778 | 40 | 0.061 | 0.053 | 0.738 |
| 13 | 0.057 | 0.053 | 0.692 | 27 | 0.065 | 0.053 | 0.783 | 41 | 0.064 | 0.053 | 0.772 |
| 14 | 0.064 | 0.053 | 0.772 | 28 | 0.064 | 0.053 | 0.773 | 42 | 0.063 | 0.053 | 0.756 |

Design or 7, 16 and 25 were not analysed because they gave the results that were considered irrational.

From the overall designed propeller, then first stage scoring is executed. The results of first stage scoring are summarized in Table 4 below,

**Table 4**. The Results of First Stage Scoring

| Design | $S_1$ | Design | $S_1$ | Design | $S_1$ | Design | $S_1$ | Design | $S_1$ | Design | $S_1$ |
|---|---|---|---|---|---|---|---|---|---|---|---|
| 1 | 0.721 | 8 | 0.782 | 15 | 0.783 | 22 | 0.692 | 29 | 0.586 | 36 | 0.760 |
| 2 | 0.766 | 9 | 0.740 | 16 | - | 23 | 0.665 | 30 | 0.733 | 37 | 0.769 |
| 3 | 0.782 | 10 | 0.755 | 17 | 0.768 | 24 | 0.733 | 31 | 0.763 | 38 | 0.748 |
| 4 | 0.705 | 11 | 0.734 | 18 | 0.782 | 25 | - | 32 | 0.637 | 39 | 0.782 |
| 5 | 0.780 | 12 | 0.776 | 19 | 0.769 | 26 | 0.778 | 33 | 0.692 | 40 | 0.665 |
| 6 | 0.769 | 13 | 0.692 | 20 | 0.355 | 27 | 0.705 | 34 | 0.782 | 41 | 0.772 |
| 7 | - | 14 | 0.772 | 21 | 0.684 | 28 | 0.696 | 35 | 0.636 | 42 | 0.756 |

From Table 4, it can be seen that 5 propeller designs that have the highest on-design performances are design 3, 8, 15, 34, and 39. Furthermore, the propeller performances of the 5 selected designs are then analyzed along a range of advance ratio from 0.147 to 0.712 using inversing Larrabee design method. The plot of the propeller performances in the form of thrust coefficient, power coefficient, and efficiency of the selected propeller designs are shown in Fig. 3 below,

IOP Conf. Series: Materials Science and Engineering **645** (2019) 012021 doi:10.1088/1757-899X/645/1/012021

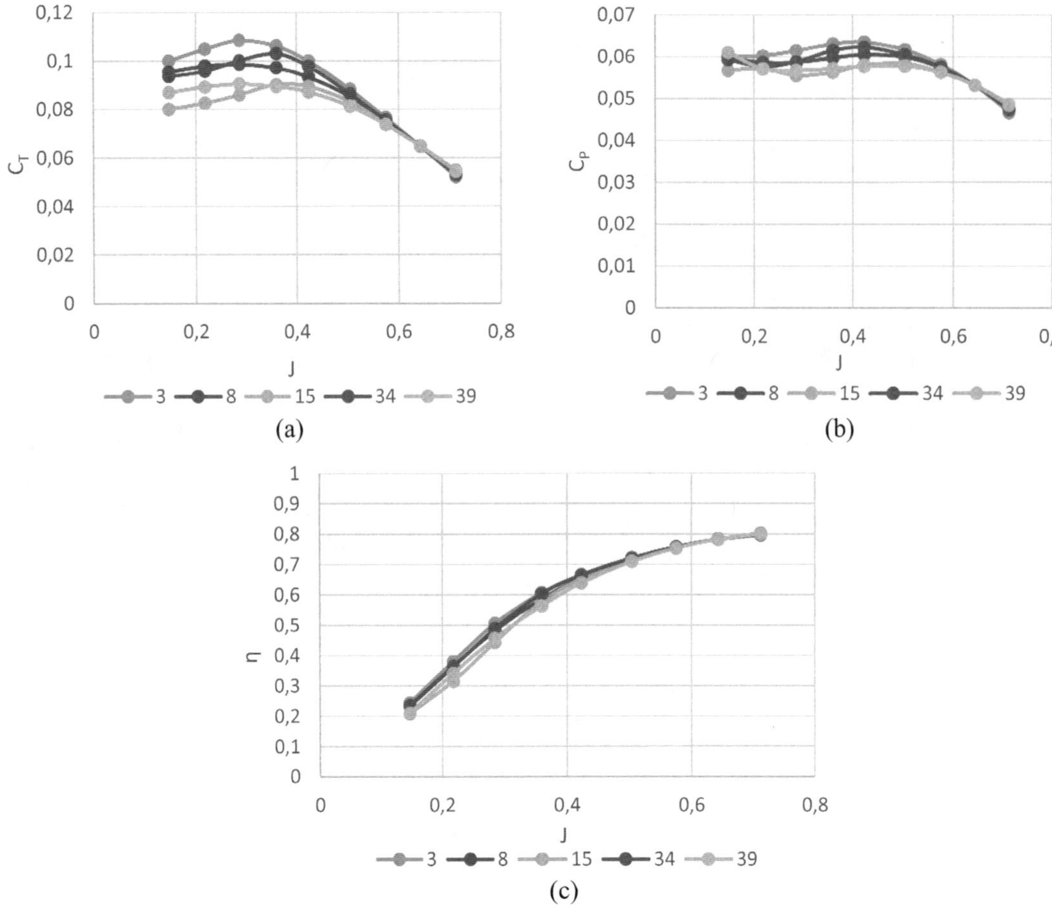


**Figure 3**. Performance of Propeller Design 3, 8, 15, 34, and 39, (a) $C_T$ vs $J$, (b) $C_P$ vs $J$, and (c) $\eta$ vs $J$

Finally, second stage scoring was carried out on 5 selected designs. The results of second stage scoring are summarized in Table 5 below,

**Table 5**. The Results of Second Stage Scoring

| Design | $S_2$ |
|---|---|
| 3 | 0.636 |
| 8 | 0.647 |
| 15 | 0.622 |
| 34 | 0.664 |
| 39 | 0.619 |

Table 5 shows that design 34 has the highest overall performances.

IOP Conf. Series: Materials Science and Engineering **645** (2019) 012021 doi:10.1088/1757-899X/645/1/012021

## 6. Analysis

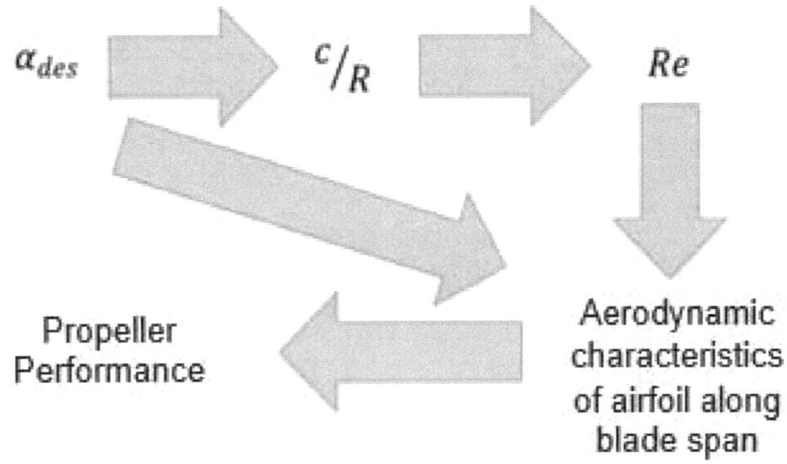


**Figure 4**. The Scheme How $\alpha_{des}$ Affects the Propellers Performance

Based on the study of $\alpha_{des}$ variations that have been carried out, it can be concluded that $\alpha_{des}$ affects the performance of the propellers with the scheme as shown in Fig. 4. It can be seen that $\alpha_{des}$ along the blade span will affect the output of the chord distribution that is generated at the propeller design stage. Then, the chord distribution will affect the Reynolds number that works along the propeller blades. The Reynolds number that works on the propeller blade affects the aerodynamic characteristics of the airfoil blades. In addition, the aerodynamic characteristics of the airfoil blades are also directly affected by the $\alpha_{des}$ distribution along the blade span. Finally, airfoil aerodynamic characteristics affect the performance of the propeller design.

Propeller that provides the best overall performance, which in this research is design 34, is a propeller that has the best aerodynamic efficiency along the advance ratio, which is affected either directly or indirectly by $\alpha_{des}$. It can be seen from Fig. 2(c) that design 34 has quadratic distribution of $\alpha_{des}$ along the blade span. But in general, $\alpha_{des}$ with quadratic distribution is not always provides the best overall performance, because the results that are obtained in this research are also influenced by other inputs, such as type of airfoil. It's difficult to get an alpha design distribution which produces the best overall propeller performance in all design cases. So, if a similar research is carried out but using different input, for example, type of airfoil, then the results obtained may be different from the results of this research.

## 7. Conclusions

This study conclude that the alpha design ($\alpha_{des}$)affects the performance of the propeller, where design 34 provides the best overall performance of the propeller. The distribution of $\alpha_{des}$ which produces the best overall propeller performances is still uncertainty as the performance are also influenced by other inputs. Further study on airfoil variations used for propeller blades is needed.

**Acknowledgments**
The authors would like to thank the Department of Aerospace Engineering and Bandung Institute of Technology for their support in these efforts.

## References

[1] Brandt, J. B., & Selig, M. S., "Propeller Performance Data at Low Reynolds Numbers", AIAA Paper 2011-1255, 2011.

[2] Merchant, M. P., & Miller, L. S., "Propeller Performance Measurement for Low Reynolds Number UAV Applications", AIAA Paper 2006-1127, 2006.

[3] Kutty, H. A., & Rajendran, P., "3D CFD Simulation and Experimental Validation of Small APC Slow Flyer Propeller Blade", Aerospace Journal, vol. 4, pp. 1-11, 2017.

[4] Kumar, J. U., Narvydas, E., & Eidukynas, D., "Reverse Engineering and CFD Analysis of Propeller", Conference: Young Researcher Conference 'Industrial Engineering 217, 2017.

[5] Brandt, J.B., Deters, R.W., Ananda, G.K., & Selig, M.S. UIUC propeller data site [Online] Available: http://m-selig.ae.illinois.edu/props/propDB.html [Accessed on 23 April 2018]

IOP Conf. Series: Materials Science and Engineering **645** (2019) 012022 doi:10.1088/1757-899X/645/1/012022

# Aviation and Aircraft Engine Emissions at Juanda International Airport

**F D Chilongola[12] and E Ahyudanari[1]**

[1] Department of Civil Engineering, Institut Teknologi Sepuluh Nopember, Sukolilo 60111, Surabaya, Indonesia
[2] Tanzania Civil Aviation Authority, Aeronautical Information Management,Tanzania

Email: ervina@ce.its.ac.id

**Abstract**. Aviation industry growth has increased continuously over the years. Air transport demand continues to increase due to the reliability, time saving and efficiency. Aviation industry has impact on environment as a source of pollutant emissions. The estimation of pollutant emissions at taxi/idle, approach, climb out and take off of Carbon dioxide ($CO_2$) and Nitrogen dioxide ($NO_x$) at Juanda International Airport Indonesia was calculated using Advanced ICAO landing and take-off (LTO) cycle method. The $CO_2$ emissions were 23.02 tons, 10.52 tons, 17.27 tons and 6.73 tons during taxi/idle, approach, climb out, take off respectively while NOx emissions were 0.03 tons, 0.03 tons, 0.12 tons and 0.06 tons for taxi/idle, approach, climb out and take-off respectively. Based on the results of the discussion it can be concluded that pollutant emissions and fuel consumption are positively related to amount of flight operation. The aircraft emissions affect the air quality around the vicinity of airport and lead to climate change.

## 1. Introduction

The Aviation industry growth has increased continuously over the years. Air transport demand continues to increase due to the reliability, time saving and efficiency. Indonesia as Asean country has been the one of fastest growing economic country in the world. According to the Asean aviation markets will be duplicated in the coming years [1]. Global aviation increasing annual at the rate of 5% in recent years and forecasted to increase [2,3]. Aviation industry has impact on environment as a sources of pollutant emissions.

The fuel burned by aircraft engines emits pollutants such carbon dioxide ($CO_2$), carbon monoxide (CO), water vapors ($H_2O$), hydrocarbons (HC) nitrous oxide ($NO_X$), Sulphur dioxide and particle matter which are the key to the environmental impact on the atmosphere. Carbon dioxide and water vapour emitted when complete fuel combustion occurs. Carbon dioxide is the greenhouse gases which has a great effect on earth climate warming [4]. In is founded that in aviation fuel the ratio of carbon to hydrogen lies within the range of 3148-3173g kg Fuel-1 [5]. [6] proposed international aviation carbon dioxide ($CO_2$) emissions will rise by more than 110% between 2005 and 2025. Global aviation's $CO_2$ emissions denoted at 22%coming from European aviation [7]. Nitrogen oxides are produced when air passes through high temperature or high-pressure combustion and nitrogen and oxygen present in the air combine to form NOx. A five tons heavier aircraft on cruising at a 2000 ft higher altitude yields in 11.2% (12.3%) more or 2.7% (2.4%) less NOx emissions (NOx intensity), respectively [8].

The emitted pollutant from aircraft engines have environmental impact on air quality in the vicinity of airport and climate changes. The health problems caused by emitted pollutants are such as premature mortality, lung function impairment, eye and respiratory tract infections [9]. Current aircraft operations at cruise altitude (about 80%) and the landing and takeoff cycle (LTO) (about 20%) contributed approximately 10,000 premature mortalities per year [10]. Aviation impacts on global climate changes due to emissions of greenhouse gases such as $CO_2$ and water vapor impact include higher average temperatures, ocean warming, ocean acidification, sea level rise, changes in precipitation, decreasing sea ice, and changes in physical and biological systems which lead to effect on airport and aircraft activities such as aircraft performance, heat damage to infrastructure, delays, cancellations, flooding of airport and access route, crossing wind changes affecting airport capacity and network disruption. The wealth loss is also the damage caused by climate changes [11].

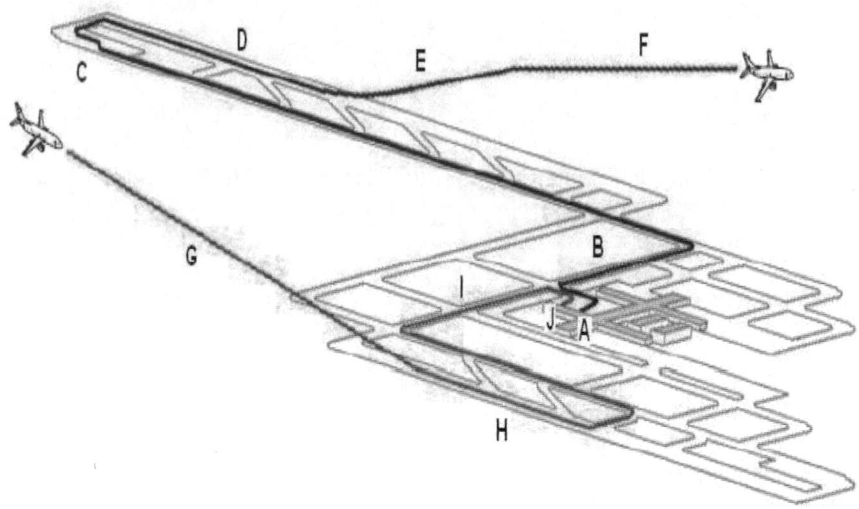


**Figure 1.** ICAO Standard landing and take-off (LTO) cycle.

There are several studies dealing on emission of pollutant from aircraft engine exhaust due to the significant growth of air traffic around the world. Other applied an advanced approach using a landing and take-off cycle (LTO) assessing aircraft emissions and pollutant trends at Nikola Tesla International Airport in Belgrade, Serbia, found that the number of LTO cycle increased by 54% which lead the increase in local air pollution, as presented on Fig. 1 [12], as well as other researcher was using three models (EDMS, LASPORT and ALAQS-AV) to calculate Zurich Airport emissions, the outputs of the comparison showed that all the models have similar global outcomes for aircraft and stationary source emissions [13]. In another study, it presents a detailed review of the global civil aviation industry for reducing $CO_2$ emissions from aircraft [14].

In order to know the extent to which action is required, it is necessary to determine the level of air pollution near the airport. Presidential Regulation Republic of Indonesia No. 61 of 2011, which aims to reduce the levels of greenhouse gases, only by 26% -46% in all areas, but more focus placed on emission produced by land transportation while by air transportation is still lacking. This paper deal with estimating the pollutant emissions of carbon dioxide ($CO_2$) and Nitrogen oxide ($NO_X$) at LTO cycles at a Juanda International Airport during peak hour.

## 2. Methods

### *2.1 Flight operation modes*

Aircraft operation can be branched into two parts landing and take-off cycle (LTO cycle) and cruise part [15]. Other researchers defined the LTO cycle as all the operations that occur when aircraft is below 3000 ft above field elevation (equivalent to 914 m) over a specific range of certifiable operating conditions [16]. The LTO cycle has four operation mode. It is starting by aircraft landing below 3000ft and then processing tax in towards airport terminal, after loading and unloading aircraft start to tax out and prepare for take-off. Then entering climb out phase until to the 3000 ft above field elevation. The schematic diagram off LTO cycle operation is shown on Fig. 1. The time spend on each mode is known as Time-in-mode (TIM) and is proportional to the amount of emissions during stage of the LTO cycle. Thrust setting and time in mode on each operation as shown on Table 1. This study will use time in mode provided by ICAO as reference because ICAO as a key holder and international accepted standard in aviation industry. However, duration of approach and take off depend on local meteorological conduction while taxing time depend on the operating procedure and the size of the airport. All activities of the aircraft occur above 1000m categorized as cruise part.

**Table 1.** Time in mode on LTO cycle ICAO (2011) [17]

| Operation Mode | Thrust Setting (%) | Time in Mode (min) |
|---|---|---|
| Taxi/idle | 100 | 26 |
| Approach | 30 | 4 |
| Climb out | 85 | 2.2 |
| Take off | 100 | 0.7 |

*2.2 Aircraft Emission calculation*

To calculate landing and take-off pollutant of emission at Juanda International Airport International Civil Aviation Organization (ICAO) methodology was used. This methodology is common and have been used in most airport all over the world due to ICAO is the key and reputable authority on aviation industry. Advanced approached was used to calculate the emission of Nitrogen dioxide (NOx) as on the following equation (1) [17].

$$E_{IJ} = \sum(TIM_{jk}) \times 60) \times (FF_{jk}/1000) \times E_{jk} \times N_J \quad (1)$$

Where;

$TIM_{jk}$: time-in-mode k for the aircraft type j. in minutes (min)

$FF_{jk}$: the fuel flow for mode k in kilograms of fuel per second (kg/s), for each engine used on the aircraft type j.

$EI_{jk}$: the emission index for pollutant i given in grams of pollutant per kilogram of fuel (g/kg fuel) in mode k (during take-off, climb, idle, and approach) for each engine used on the aircraft type j,

$NE_j$: the number of engines used on the aircraft type j.

ICAO does not set standard for $CO_2$ .hence using emission indices provided by [18] as 3155 g.kg$^{-1}$. Since the emission indices based on fuel composition, the following expression can be used to calculate emission of carbon dioxide as below.

$$E_j = \sum(TIM_j \times 60) \times ER_j \times NE_J \quad (2)$$

Where;

TIMj: time-in-mode in minutes (min) for the aircraft type j,

ERj: the emission rate of total $CO_2$ emitted per second per the mode (g/s); for aircraft j:

ERj = 3155×FFj,

FFj: the fuel consumption per mode for each engine on the aircraft type j (kg/s).

*2.3 Aircraft movement and Engine*

Aircraft usual fly under two types of flight rules namely, Instrument flight rules (IFR), aircraft flying under IFR are typically assigned altitudes of odd/even thousand feet (example 10000 ft, 20000 ft, etc.) Above Mean Sea Level (AMSL) and Visual flight rules (VFR), aircraft flying using VFR assigned odd/even thousands of feet plus 500 ft. of altitude. About 60% to 80% of emissions coming from IFR flights while VFR flights represent less than 5% of fuel consumption and pollutant emissions caused by air traffic [17, 18].

Aircraft engine are divided into three different category turbofan (jet), turboprop and piston engine. According to previous result, the turboprop and turbofan engines are powered by aviation kerosene also known as jet fuel while piston engines are fueled by aviation gasoline also namely as avigas [17]. Turbofan engines, the main reference is made to bypass ratio which is defined as the ratio of the mass airflow through the fan to the mass airflow through the core of the engine or the turbojet portion In civil aviation for a good fuel efficiency and low noise, high-bypass ratios are preferred [19].

Emission Index is defined as the mass of the pollutant emitted per unit mass of fuel burned for a specified engine. The emission indices depend on conditions and measurements procedure during aircraft engine certifications. Emission indices vary according to type of an engine mounted on an aircraft, life expectancy (age) of aircraft and metrological conditions [20]. This research used flight data from www.flightradar24.com to obtained schedule movement at Juanda International which contain aircraft types, airline, date of operation, origin and destination. On peak hour on 21/09/2018. Official aircraft engine manufactures were used to obtain typical engine combination mounted on the different types of aircraft. The ICAO Emission indices databank 2018 was used for fuel flow factor and emission indices of each engine. This paper assesses only turbofan engine.

## 3. Results and Discussions

The fuel consumptions during peak hour at Juanda international Airport were estimated around 7307 kg, 3338kg, 5482kg and 2135kg for idle/taxi, approach, climb out and take off respectively. The fuel consumption is higher during taxi/idle due the time aircraft spend during taxi in and taxi out. Results shown on Fig. 2 to Fig. 6 and Table 2 and Table 3.

Figure 2 shows the types of aircraft at Juanda International airport during peak hour. Boeing 737-800 (B738) contribute 33% of all aircraft movements followed by Airbus 320-214 (A320) which made up 24% of all movements. The pollutant emission of carbon dioxide $CO_2$ during taxi/idle was 40% of the total amount of LTO cycle, highest during taxi/idle because the emission of carbon dioxide depends solely on the amount of fuel consumption and during taxi/idle the amount of fuel consumption was high. The distribution of amount of carbon dioxide emission during tax//idle/ approach. Climb out and take off shown on Fig. 3. As pollutant emissions are positively correlated with the fuel consumption [21].

Figure 4 depicts the Nitrogen dioxide emission which is 48% at climb out and 25% during take-off. The NOx emissions are high when the power setting and temperature increase. Hence during climb out and take off the emissions were high. The same results shown by others, where taxi/idle, approach, climb out and take-off was 49%, 13%,13% and 25% respectively [22].

Boeing 737-800 (B738) contribute maximum amount of 33% while B733 minimum at 4% of total $C0_2$ at LTO cycle. This is because B738 is the most frequently operated during peak hour as seen on Figure 5. The research results give a picture on how much emission has been loaded into the air from aircraft movement at Juanda International airport.

**Table 2.** Fuel consumption during peak hour.

| Mode | Fuel Consumption (kg) |
|---|---|
| Taxi/Idle | 7307 |
| Approach | 3338 |
| Climb out | 5482 |
| Takeoff | 2135 |
| TOTAL | 18263 |

IOP Conf. Series: Materials Science and Engineering **645** (2019) 012022 doi:10.1088/1757-899X/645/1/012022

**Table 3.** Pollutant emission at Juanda International airport.

| Mode | $CO_2$ Emission (tons) | $NO_X$ Emission(tons) |
|---|---|---|
| Taxi/Idle | 23.02 | 0.03354 |
| Approach | 10.52 | 0.03448 |
| Climb out | 17.27 | 0.12248 |
| Takeoff | 6.73 | 0.06209 |
| TOTAL | 57.53 | 0.25260 |

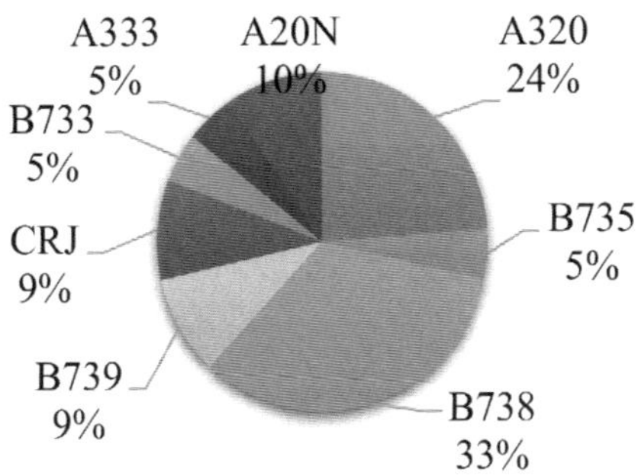


**Figure 2.** Aircraft movement distributions.

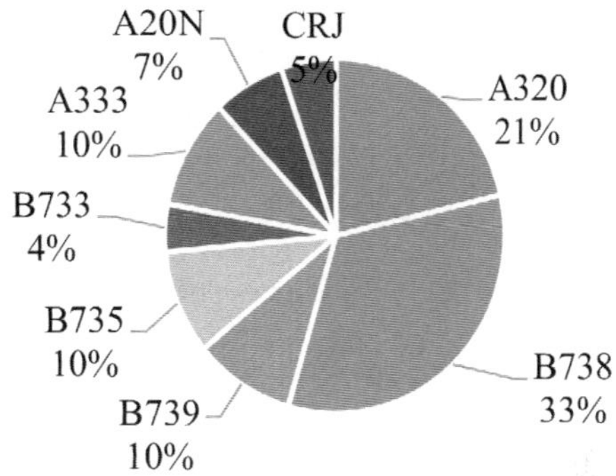


**Figure 5.** $CO_2$ emissions by aircraft type.

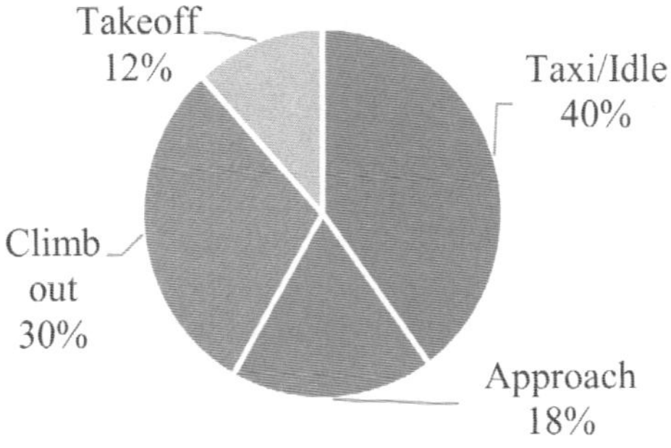


**Figure 3.** $CO_2$ emissions distribution.

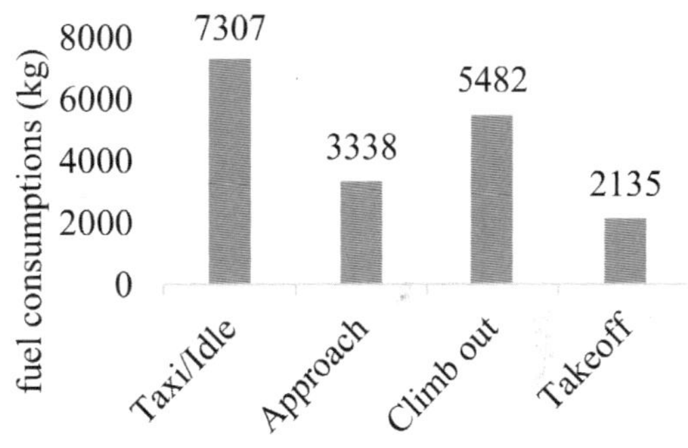


**Figure 6**. Total fuel consumptions during peak hour.

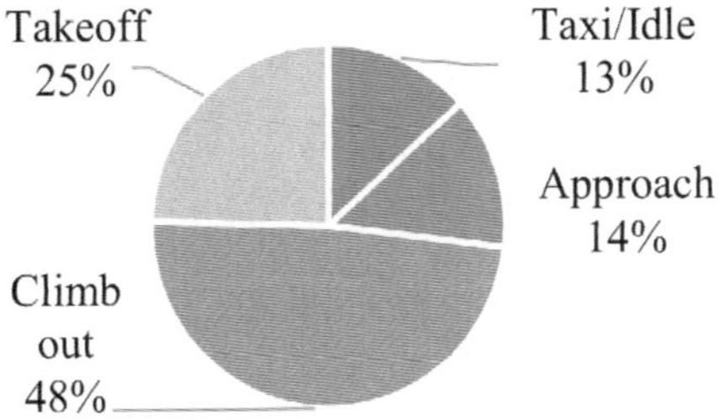


**Figure 4**. $NO_X$ emissions distribution.

## 4. Conclusion

This paper estimates the amount of fuel consumption and pollutant emission of $CO_2$ and $NO_X$ during LTO cycle by using ICAO advanced approach methodology. The estimation of total fuel consumption during taxi/idle, approach, climb out and takeoff was 7307 kg, 3338 kg, 5482 kg and 2135 kg respectively. The high amount of fuel consumption was during taxi/idle therefore the time spent in each mode is positively correlated to amount of fuel consumptions. The $CO_2$ emissions were 23.02 tons, 10.52 tons, 17.27 tons and 6.73 tons during taxi/idle, approach, climb out, take off respectively while NOx emissions were 0.03354 tons, 0.03448 tons, 0.12248 tons and 0.06209 tons for taxi/idle, approach, climb out and take off respectively. The aircraft emissions affect the airport air quality that may lead to environmental climate changes.

**Acknowledgement**
This paper is part of Research Grant under EPI-UNET Scheme Year 2018.

## References

[1] Hooper P 2005 The environment for Southeast Asia's new and evolving airlines *J. Air Transp. Manag.* **11** 335–47
[2] Airbus 2012 *Global Market Forecast 2012-2031*
[3] Boeing 2013 *Current Market Outlook 2013-2032. Market Analysis, Boeing Commercial Airplanes, Seattle*
[4] IPCC 1999 *Aviation and the Global Atmosphere. Summary for Policymakers. IPCC Working Groups I and III, p. 23.*
[5] Hileman J I, Donohoo P E and Stratton R W 2010 Energy content and alternative jet fuel viability *J. Propuls. Power* **26** 1184–96
[6] Macintosh A and Wallace L 2009 International aviation emissions to 2025: Can emissions be stabilised without restricting demand? *Energy Policy* **37** 264–73
[7] EEA 2014 Transport and environment reporting mechanism 2014
[8] Turgut E T and Usanmaz O 2017 An assessment of cruise NOxemissions of short-haul commercial flights *Atmos. Environ.* **171** 191–204
[9] Neufville R de, Odoni A R, Belobaba P and Reynolds T 2013 *Airport systems: planning, design, and management* (McGraw-Hill Education LLC)
[10] Barrett S R H, Britter R E and Waitz I A 2010 Global mortality attributable to aircraft cruise emissions *Environ. Sci. Technol.* **44** 7736–42
[11] Fuglestvedt J S, Shine K P, Berntsen T, Cook J, Lee D S, Stenke A, Skeie R B, Velders G J M and Waitz I A 2010 Transport impacts on atmosphere and climate: Metrics *Atmos. Environ.* **44** 4648–77
[12] Vujović D and Todorović N 2017 An assessment of pollutant emissions due to air traffic at Nikola Tesla International Airport, Belgrade, and the link between local air quality and weather types *Transp. Res. Part D Transp. Environ.* **56** 85–94
[13] Celikel A, Duchene N, Fuller I, Fleuti E and Hofmann P 2005 Airport local air quality modeling: Zurich airport emissions inventory using three methodologies *EUROCONTROL Air Transport Research Society (ATRS) Conference, Rio de Janeiro, Brazil*
[14] Grote M, Williams I and Preston J 2014 Direct carbon dioxide emissions from civil aircraft *Atmos. Environ.* **95** 214–24
[15] EEA/EMEP 2009 Emission Inventory Guidebook: 2009
[16] ICAO 1993 International Standards and Recommended Practices, Environmental Protection Annex 16. Volume II Aircraft Engine Emissions. second ed.
[17] International Civil Aviation Organization (ICAO) 2011 *Airport air quality manual* vol 1
[18] Kalivoda M T and Kudrna M 1997 Methodologies for estimating emissions from air traffic *Oct. MEET Proj. Contract No. ST-96-SC.204.*

[19] Masiol M and Harrison R M 2014 Aircraft engine exhaust emissions and other airport-related contributions to ambient air pollution: A review *Atmos. Environ.* **95** 409–55

[20] Zaporozhets O and Synylo K 2017 Improvements on aircraft engine emission and emission inventory asesessment inside the airport area *Adv. Energy Technol. Aviat.* **140** 1350–7

[21] Fan W, Sun Y, Zhu T and Wen Y 2012 Emissions of HC, CO, NOx, CO2, and SO2 from civil aviation in China in 2010 *Atmos. Environ.* **56** 52–7

[22] Yılmaz İ 2017 Emissions from passenger aircrafts at Kayseri Airport, Turkey *J. Air Transp. Manag.* **58** 176–82

# On stabilizing hydraulic motion experiencing Stribeck friction via PI controller: Circle criterion approach

**T P Tumbuan[1], Indrawanto[2], I P Nurprasetio[1], Z Abidin[1]**

[1] Mechanical Design, FMAE, ITB, Bandung, Indonesia
[2] Mechanical Production Engineering, FMAE, ITB, Bandung, Indonesia

Email: tobias.tumbuan@gmail.com

**Abstract.** A Friction is inevitable in a hydraulic actuator due to the presence of seal. Unfortunately, the existence of friction makes the system prone to limit cycle oscillation as PI controller is applied. This research investigates the maximum integrator constant of PI controller that can be introduced to the system while maintaining global asymptotic stability in Lyapunov sense. This condition is achieved as the system complies with circle criterion. In this research, a simplified hydraulic system which behaves linearly is adopted. Thus, the only source of nonlinearity is the Stribeck friction. The main contribution of this research is a formulation of stable PI strategy that is free from the effect of inertia. The PI strategy is suitable for a hydraulic testing machine of which the mass of device under the test is varying.

**Keywords:** Stribeck friction, PI controller, hydraulic, limit cycle oscillation, circle criterion

## 1. Introduction

Hydraulic based position control is applied in the various industries for its capability to handle high load. To ensure the stability and performance of the control system, the old fashioned PID is still employed. Unfortunately, the expected performance cannot be achieved effortlessly as friction is inevitable in hydraulic actuator. This friction is generated by the seal of hydraulic actuator to reduce leakage during the energy transformation from fluid to mechanical. The friction makes the system liable to limit cycle oscillation and hunt the final position. This condition diminishes the accuracy of hydraulic positioning control.

There are a lot of friction models that have been developed to capture various friction behaviors. The comparison of this model is well investigated by Liu [1]. However, based on the experiment led by Bistray [2], Stribeck friction fit to describe the friction of Hydraulic actuator sealing system. Regardless the cause of friction, Marton [3] and Qu [4] also found the Stribeck behavior on the hydraulic cylinder actuator. The Stribeck friction model is characterized by strong nonlinearities due to the presence of sign function in pre-sliding regime. The more relax nonlinearities during the pre-sliding regime is given by Sin [5] who employed what so called Stribeck friction with threshold. On this model, the pre-sliding regime is characterized by a finite slope but not hysterical.

A breakthrough analysis about friction that causes limit cycle in mechanical position control scheme is developed by Hensen [6],[7]. The result of Hensen demonstrated the significance of integrator gain of PID that gives rise limit cycle of the system. In accordance to the result of Hensen, Maneetham [8] applied PD to stabilize position control of high speed servo system. Moreover, advance control algorithm such as sliding mode control was also applied to deal with limit cycle due to friction [9]. Unfortunately, the application of PD controller causes a steady state error [10] while the advanced controller is not applicable in industrial world. Thus, some researcher devoted their research to understand the behavior of limit cycle due to friction in position control scheme.

The behavior of limit cycle in hydraulic position control is investigated by Chin [11] and Yanjay [12]. Both Chin and Yanjay investigated the existence of limit cycle via describing function method. While Chin only focus on the effect of Stribeck friction to the existence of limit cycle, Yanjay analyzed more in depth to the interaction between transport lag and Stribeck friction. However, both Chin and Yanjay do not give any criterion to avoid the limit cycle oscillation.

The limit cycle problem due to friction can also be addressed by dissipating all of the energy of the system by means fulfilling global asymptotic stability in Lypunov sense. This approach is done by Lee [13] who developed Lypunov energy like function of the system and found the PID criterion that will stabilize it. A more straight forward method to reach global asymptotic stability is to ensure the system comply with absolute stability. The application of absolute stability is suggested by Golestani [14]. Golestani developed a stable PID strategy based on circle criterion for a simple mechanical system.

This research is intended to maintain the research path of Chin [11] and Yanjay [12]. In this research, a PI criterion that will stabilize hydraulic position control in the presence of Stribeck friction is well developed. The Stribeck model applied on this research is Stribeck friction with the threshold applied by Sin [7]. The PI strategy is found as the system conforms to circle criterion that will make all of the energy of the system be dissipated. Having this approach, the hydraulic system needs to be simplified so that it behaves as a linear system. Hence, the only non-linearity present on the system is the Stribeck friction. The analysis shows that the maximum integrator gain of PI controller depends on the Stribeck properties during the pre-sliding behavior (stick behaviours). In addition, the developed PI strategy is independent of inertia. Thus, the PI guideline is very convenient to be applied to universal hydraulic testing of which the mass of device under the test is varying.

## 2. Hydraulic System and Stribeck Friction

Hydraulic system is a highly nonlinear system. This nonlinearity is generated by pressure-volume formulation of the fluid system. In order to simplify this formulation, the pressure-volume mapping is linearized. This research follows the linearization of symmetric hydraulic cylinder developed by other [15] who simplified the hydraulic system into second order system. The final block diagram of the symmetric hydraulic cylinder with PI controller is shown in Figure 1 (a). In the figure 1, $x_{Ref}$, $x_{Act}$, $K_P$, and $K_I$ represents reference position ($x_{Ref}$), actual position ($x_{Act}$), proportional gain ($K_P$) and integrator gain ($K_I$) of the controller. Moreover, $M$, $A_P$, $B$, $K_Q$, $K_C$, and $K_{c\beta}$ represent mass ($M$), area of the cylinder ($A_P$), viscous friction ($B$), gain coefficient ($K_Q$), flow pressure coefficient ($K_C$), and compress flow coefficient ($K_{c\beta}$) respectively. The compress flow coefficient is determined by bulk modulus ($\beta$) of the fluid and the volume of the chamber ($V_T$) as it is shown in Eq. 1.

$$K_{C\beta} = \frac{V_T}{4\beta} \tag{1}$$

In order to make the system can be subjected to the circle criterion, it has to be represented in a Lure form. In the Lure representation, the system which behaves linear is separated from its nonlinear feedback. Moreover, the Lure system has no input and thus it will reveal the true behavior of the system. The lure system of the system above is shown in Figure 1 (b). In this Lure representation, the linear system is denoted by $G(s)$ of which the feedback is Stribeck nonlinear friction. Having an elaboration from the Lure representation above, the formulation of $G(s)$ can be restated as it is shown in Eq. 2. In the light of Hurwitz stability, $G(s)$ will be stable if $K_I$ is less than $K_{I\text{-Lin}}$ and $K_P$ is less than $K_{P\text{-Lin}}$. The requirement of Hurwitz stability for the following system is shown in equation (3).

$$G(s) = \frac{s^2(K_C + K_{V\beta}s)}{MK_{V\beta}s^4 + (BK_{V\beta} + MK_C)s^3 + (BK_C + A_P{}^2)s^2 + A_P K_Q K_P s + A_P K_Q K_I} \tag{2}$$

$$K_I < K_{I\text{-Lin}} = \frac{(BK_C + A_P{}^2)K_P}{(BK_{V\beta} + MK_C)} - MK_{V\beta}\frac{A_P K_Q K_P{}^2}{(BK_{V\beta} + MK_C)^2} \;\wedge\; K_P < K_{P\text{-Lin}} = \frac{(BK_{V\beta} + MK_C)(BK_C + A_P{}^2)}{A_P K_Q M K_{V\beta}} \tag{3}$$

The formulation of Stribeck friction is the combination of coulomb friction, viscous friction and Stribeck phenomenon inherently. However, as the viscous friction possess linear behavior, it is covered by the linear system G(s). For that reason, the Stribeck formulation employed in this research is only the combination of coulomb friction and the Stribeck phenomenon as it is shown in Eq. 4. There are two regimes defined in this Stribeck formulation. Those are the pre-sliding regime where $|v|<\beta$ and slipping regime where $|v| \geq \varepsilon$. In this formulation, $F_f$, $F_s$, and $F_c$ represent friction force ($F_f$), static friction force ($F_s$), and coulomb friction ($F_c$) respectively. Meanwhile, $v$, $v_s$, $\varepsilon$, and n stand for velocity ($v$), Stribeck velocity coefficient ($v_s$), and Stribeck threshold ($\varepsilon$), Stribeck power coefficient ($n$) respectively.

$$F_f(v)\begin{cases} = \dfrac{F_s}{\varepsilon}v & \text{for } |v| < \varepsilon \\ = \left\{ F_c + (F_s - F_c)\exp\left[-\left(\dfrac{|v|-\varepsilon}{v_s}\right)^n\right]\right\}\operatorname{sign}(v) & \text{for } |v| \geq \varepsilon \end{cases} \quad (4)$$

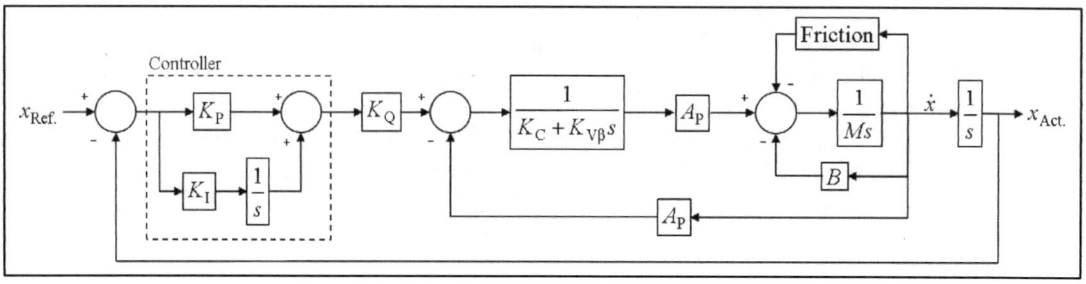


(a)

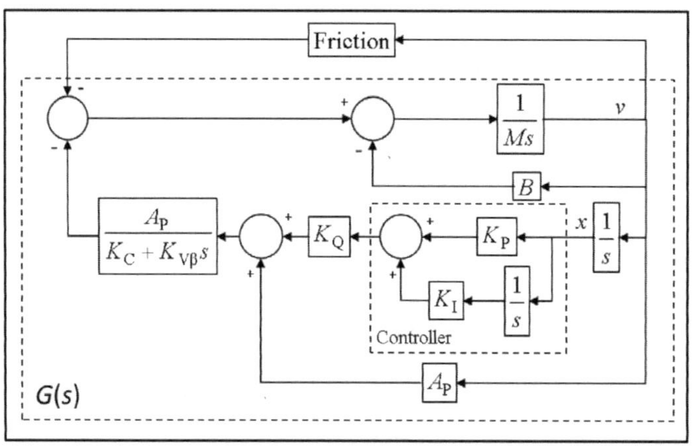


(b)

**Figure 1.** Position control scheme of symmetric hydraulic cylinder via PID. (a) Simplified block diagram (reproduced from [15]). The symmetric hydraulic cylinder system is simplified into the second order system. (b) The Lure representation of the corresponding blocking diagram. In this Lure representation, the linear part of the system is denoted by $G(s)$ which is fed-back by a friction nonlinear function.

## 3. Circle Stability

Circle stability is an absolute stability criterion built under frequency domain analysis and sector nonlinearity. Observing Eq. 4, the Stribeck nonlinearity sector is bounded between $m_1$ and $m_2$ where $m_1 = F_s/\varepsilon$ and $m_2 = 0$. Based on circle criterion, the stability would be achieved as $G(s)$ fulfils Hurwitz and the Nyquist plot of the linear system $G(s)$ does not cross the line Real = $-\ \varepsilon/F_s$. The mathematical formulation of the circle criterion for this case is shown in Eq. 5. In this formulation, imaginary unit is denoted by i.

$$\mathrm{Real}[G(\mathrm{i}\omega)] = \mathrm{Real}\left[\frac{-\omega^2(K_\mathrm{C} + \mathrm{i}K_{\mathrm{V\beta}}\omega)}{MK_{\mathrm{V\beta}}\omega^4 - \mathrm{i}(BK_{\mathrm{V\beta}} + MK_\mathrm{C})\omega^3 - (BK_\mathrm{C} + A_\mathrm{P}{}^2)\omega^2 + \mathrm{i}A_\mathrm{P}K_\mathrm{Q}K_\mathrm{P}\omega + A_\mathrm{P}K_\mathrm{Q}K_\mathrm{I}}\right] > -\frac{\varepsilon}{F_\mathrm{s}}\forall\omega \in \mathrm{Real} \quad (5)$$

Having some algebra, the above equation leads to

$$\frac{-\omega^2\left\{K_\mathrm{C}\left[MK_{\mathrm{V\beta}}\omega^4 - (BK_\mathrm{C} + A_\mathrm{P}{}^2)\omega^2 + A_\mathrm{P}K_\mathrm{Q}K_\mathrm{I}\right] - K_{\mathrm{V\beta}}\omega\left[(BK_{\mathrm{V\beta}} + MK_\mathrm{C})\omega^3 - A_\mathrm{P}K_\mathrm{Q}K_\mathrm{P}\omega\right]\right\}}{\left[MK_{\mathrm{V\beta}}\omega^4 - (BK_\mathrm{C} + A_\mathrm{P}{}^2)\omega^2 + A_\mathrm{P}K_\mathrm{Q}K_\mathrm{I}\right]^2 + \left[(BK_{\mathrm{V\beta}} + MK_\mathrm{C})\omega^3 - A_\mathrm{P}K_\mathrm{Q}K_\mathrm{P}\omega\right]^2} > -\frac{\varepsilon}{F_\mathrm{s}}\forall\omega \in \mathrm{Real} \quad (6)$$

To simplify equation (5), a new variable, $p$, is defined as $p = \omega^2$ so that equation (6) can be represented as follow.

$$C_4p^4 + C_3p^3 + C_2p^2 + C_1p^1 + C_0 > 0 \forall\ p \in \mathrm{Re}\ ;\ p > 0 \quad (7)$$

Where

$$C_4 = \frac{\varepsilon}{F_\mathrm{s}}(MK_{\mathrm{V\beta}})^2$$

$$C_3 = \frac{\varepsilon}{F_\mathrm{s}}\left[MK_{\mathrm{V\beta}}(BK_\mathrm{C} + A_\mathrm{P}{}^2) + (BK_{\mathrm{V\beta}} + BK_\mathrm{C})^2\right] + BK_{\mathrm{V\beta}}{}^2$$

$$C_2 = \frac{\varepsilon}{F_\mathrm{s}}\left[(BK_\mathrm{C} + A_\mathrm{P}{}^2)^2 + MK_{\mathrm{V\beta}}A_\mathrm{P}K_\mathrm{Q}K_\mathrm{I} - 2(K_{\mathrm{V\beta}} + K_\mathrm{C})BA_\mathrm{P}K_\mathrm{Q}K_\mathrm{P}\right] + K_\mathrm{C}(BK_\mathrm{C} + A_\mathrm{P}{}^2) - K_{\mathrm{V\beta}}A_\mathrm{P}K_\mathrm{Q}K_\mathrm{P}$$

$$C_1 = \frac{\varepsilon}{F_\mathrm{s}}\left[(BK_\mathrm{C} + A_\mathrm{P}{}^2)A_\mathrm{P}K_\mathrm{Q}K_\mathrm{I} + (A_\mathrm{P}K_\mathrm{Q}K_\mathrm{P})^2\right] - K_\mathrm{C}A_\mathrm{P}K_\mathrm{Q}K_\mathrm{I}$$

$$C_0 = \frac{\varepsilon}{F_\mathrm{s}}(A_\mathrm{P}K_\mathrm{Q}K_\mathrm{I})^2$$

The easiest solution of inequality shown in Eq. (6) is $C_k > 0$; k=0, 1, …, 4. By the observation, the value of $C_1$, $C_3$, and $C_4$ are positive. Hence, the boundary of Eq. 6 lays on the value of $C_2$ and $C_1$. The simplification of $C_2 > 0$ leads to Eq. (8) while the simplification of $C_1 > 0$ is shown in Eq. (9)

$$K_\mathrm{P} < K_{\mathrm{P\text{-}Max}} = \frac{K_\mathrm{C}(BK_\mathrm{C} + A_\mathrm{P}{}^2)}{K_{\mathrm{V\beta}}A_\mathrm{P}K_\mathrm{Q}} \quad (8)$$

$$K_\mathrm{I} < K_{\mathrm{I\text{-}Max}} = \frac{\varepsilon}{F_\mathrm{s}}\left(\frac{A_\mathrm{P}K_\mathrm{Q}}{K_\mathrm{C}}\right)K_\mathrm{P}{}^2 \quad (9)$$

Based on the derivation above, there is a maximum value of $K_\mathrm{P}$ and $K_\mathrm{I}$ to guarantee the system stability. The maximum value of $K_\mathrm{P}$ is independent of friction. It only depends on the properties of the hydraulic actuator and fluid properties. In contrast, the maximum value of $K_\mathrm{I}$ depends on the properties of friction around the pre-sliding regime ($\varepsilon/F_\mathrm{s}$) and the properties of hydraulic actuator itself. This analysis also shows that the maximum $K_\mathrm{P}$ and $K_\mathrm{I}$ are independent from the inertia of the system ($M$). It has to be noted that equation (8) and equation (9) only work given that $G(s)$ is stable in linear manner.

## 4. Illustrative example

To give a practical understanding about the strength of PI strategy above, an illustrative example is made. The actuator hydraulic as well as its friction characteristic is taken from reference [3]. Meanwhile, the valve property is extracted from Moog D765 series specification. These parameters are shown in Table 1. It has to be noted that the output of controller is assumed to be electrical current that will directly drive the torque motor of hydraulic servo. For this reason, the unit of proportional gain ($K_\mathrm{P}$) is mA/m and the unit of integrator gain ($K_\mathrm{I}$) is mA/(m.s).

There are six cases which will be evaluated on this section. Case 1-3 shows the application of equation (9) which claims that the maximum integrator gain can be applied to the system is proportional to the quadrat of $K_\mathrm{P}$. In case 1-3, the value of $K_\mathrm{P}$ varies from 50 to 500 and the mass is 20 kg. In these systems, the $K_\mathrm{I}$ is set to the value of $K_{\mathrm{I\text{-}Max}}$ per equation (9). Thus, those systems are predicted to be stable. Case 3-4 shows the practical aspect of the analysis that also claims that the maximum integrator gain is independent on the inertia of the system. In case 3-4 the mass of system varies from 100 kg to 400 kg. Meanwhile, the value of $K_\mathrm{P}$ is set to 500 and $K_\mathrm{I}$ is set to $K_{\mathrm{I\text{-}Max}}$ which is about 675.34. Thus, those systems will be stable. On the contrary in the last case, the value of $K_I$ is set to 0.8 $K_{I\text{-}Lin}$ which is computed without considering the friction nonlinearity. As the system violates the circle criterion, the system is predicted to be unstable (presented in Table 1 and Table 2).

**Table 1.** Simulation Parameter

| Parameter | Unit | Value |
|---|---|---|
| $A_P$ | $cm^2$ | 6.6 |
| $V_T$ | $cm^3$ | 177.4 |
| $\beta$ | MPa | 689.0 |
| $B$ | N/(m/s) | 1,219.1 |
| $K_Q$ | $cm^3$/s/mA | 29.5 |
| $K_C$ | $cm^3$/s/MPa | 49.2 |
| $F_s$ | N | 1,015.4 |
| $F_c$ | N | 513.2 |
| $v_s$ | mm/s | 69.6 |
| E | mm/s | 6.9 |
| N | - | 1.000 |

**Table 2.** Simulation Parameter

| Case | $K_P$ (mA/m) | Mass (kg) | $K_{I\text{-}set}$ [mA/(m.s)] | Prediction |
|---|---|---|---|---|
| 1 | 50 | 100 | 6.75 | Stable |
| 2 | 160 | 100 | 69.15 | Stable |
| 3 | 500 | 100 | 675.34 | Stable |
| 4 | 500 | 200 | 675.34 | Stable |
| 5 | 500 | 400 | 675.34 | Stable |
| 6 | 160 | 100 | 7930.93 | Unstable |

The simulation result of those cases are shown in Fig. 2. Figure 2 (a) shows the result of cases 1-3. The simulation result shows that those systems are stable with different rate of rise time. These results exhibit that a high value of $K_I$ can be employed when the system has a high value of $K_P$. Figure 2 (b) presents the result of cases 3-4. As it is shown in figure 2 (b) those systems are stable. These results demonstrate that the developed PI strategy is independent on the mass of the system. Figure 2 (c) give an example when the $K_{I\text{-}set}$ is higher than the proposed $K_{I\text{-}Max}$. Thus, the system is trapped in limit cycle oscillation as it is presented in figure 2(c).

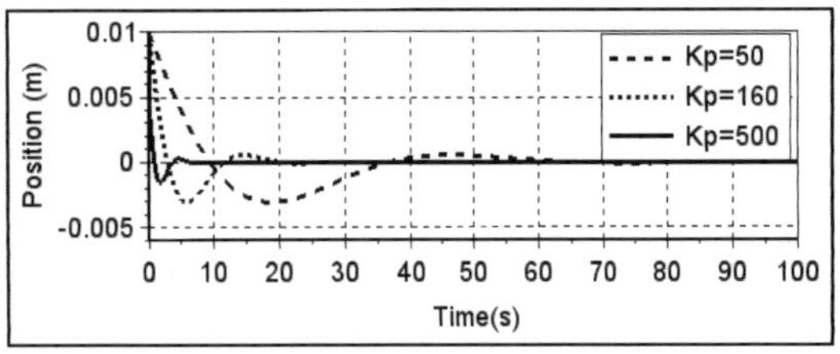


(a)

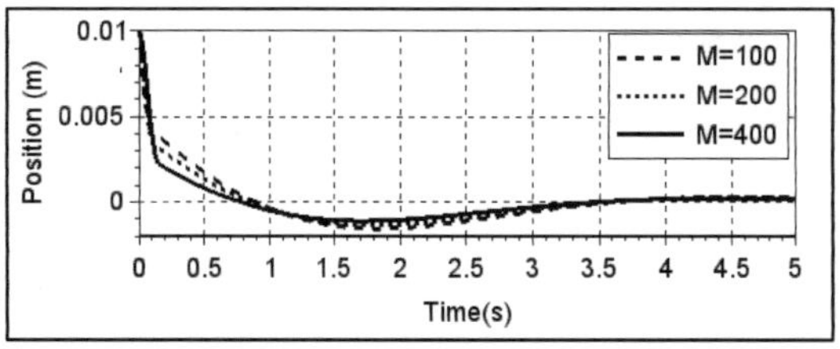


(b)

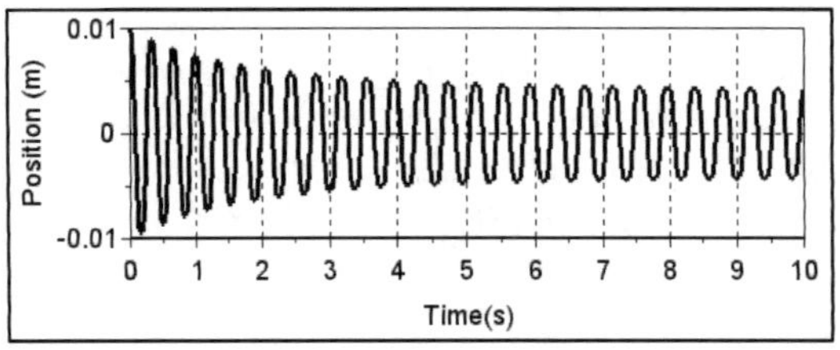


(c)

**Figure 2.** Simulation result (a) Simulation result for M=100kg with different value of $K_P$. In these cases, the $K_I$ is set to $K_{I\text{-}Max}$ such that the system fulfils circle stability and achieves stability. (b) Simulation result for $K_P$ = 500 and $K_I$ = 675.34 with different value of M. These cases show that the PI strategy is independent on the inertia of the system. (c) Simulation result for $K_P$=160; M=100 and $K_I$ is not higher than it should be so that the system is trapped in limit cycle oscillation.

## 5. Conclusion

The PI strategy has been developed to make hydraulic system achieve stability in the presence of Stribeck friction. The strategy is built under the rule of circle criterion. The analysis shows that maximum value of $K_I$ relies upon the properties of friction in the pre-sliding regime. Moreover, the maximum $K_I$ can be employed in the system that is proportional to the quadrat of $K_P$ applied. The benefit of the developed PI strategy is that it is independent on the inertia of the system. The simulation confirms this phenomenon. Unfortunately the PI strategy developed here is not the limit of instability. Thus, violating the proposed PI strategy doesn't necessarily mean that the system would be unstable. This fact is the weakness point of the approach applied in this research and might be an interesting topic to study for the future research. Moreover, this research can be followed by having an optimal control of PID for DC motor system.

**Acknowledgments**
This work is supported by ITB under P3MI research grant 2017.

**References**

[1] Y F Liu, J Li, Z M Zhang, X H Hu and W J Zhang 2015 Experimental comparison of five friction models on the same test bed of the micro stick slip motion Mechanical Sciences **6** pp 15-28

[2] Sandor B and Balku 1995 Tribology of hydraulic seals for alternating motion Periodica Polytechnica Ser. Mech. Eng. **39** pp 225-246

[3] Lorinc M, Szabolcs F and Nariman S 2011 A practical method for friction identification in hydraulic actuators Mechatronics **21** pp 350-356

[4] Sheng-guan Q, Yong D, Rong Y, Li-man H and Xiao-qiang L 2018 Study on the friction characteristics of cylinder based on Stribeck friction model Proc. of Int. Conf. on Power, Energy and Environmental Engineering pp 309-315

[5] Aung M T S, Zainal A and Hideki Y 2011 Computer simulation of unstable oscillations in a class of mechatronic systems Int. Journal of the Computer, the Internet and Management **1** pp 91.1-91.7

[6] Ron H A H, Marinus J G van de M 2002 Friction induced hunting limit cycle: An event mapping approach Proc. Of the American Control Conference pp 2267-2272

[7] Ron H A H, Marinus J G van de M and Maarten S 2003 Friction induced hunting limit cycles: A comparison between the LuGre and switch friction model Automatica **39** 2131-2137

[8] Dechrit M and Nitin A 2010 Modeling, simulation and control of high speed nonlinear hydraulic servo system World Journal of Modelling and Simulation **6** no 1 pp 27-39

[9] Indrawanto and T X Bo 2011 Sliding mode control of a single rigid hydraulically actuated manipulator Int. Journal of Mechanical and Mechatronics Engineering **11** no 5

[10] Devi P, Hengk N and Nathan van de W 2007 Analysis of under compensation and overcompensation of friction in 1 DOF mechanical systems Automatica 43 pp 1387-1394

[11] Ying-jeh H and Yuan-jay W 2002 Limit cycle analysis for electro-hydraulic system with friction Journal of the Chinese Institute of Engineers **25** issue 3 pp 277-295

[12] Wang Y, Huang J, Zhang, Y and Zhuang F 2008 Analysis and quenching of limit cycles of electro-hydraulic servo valve control system with friction and interval transport lag Proc. Of the 27th Chinese Control Conference pp 212-216

[13] Jung H L 2002 A new PID controller with Lyapunov stability for regulation servo systems Journal of IKEEE **13** no 1 pp 11-19

[14] Mehdi G and Iman M 2015 PID guidance law design using short time stability approach Aerospace Science and Technology **43** 71-76

[15] Gang S, Guang-ming L V, Zheng-mao Y, Da-cheng C and Jun-wei H 2011 Feed-forward inverse control for transient waveform replication on electro-hydraulic shaking table Journal of Vibration and Control **18**(10) pp 1471-1493.